# Classics in Psychology
# 1855–1914

## A Collection of Key Works

Thoemmes / Maruzen

# Classics in Psychology, 1855–1914
## A Collection of Key Works
Edited by **Robert H. Wozniak,** *Bryn Mawr College*

Robert H. Wozniak, *Guide to Classics in Psychology, 1855–1914* (1999)

1. A. Bain, *The Senses and the Intellect* (1855)
2. H. Spencer, *The Principles of Psychology* (1855)
3. A. Bain, *The Emotions and the Will* (1859)

4/5. G. T. Fechner, *Elemente der Psychophysik* (1860)

6. F. Galton, *Hereditary Genius* (1869)
7. H. Maudsley, *Body and Mind* (1870)
8. H. Taine, *On Intelligence* [1870] (1871)
9. H. L. F. Helmholtz, *On the Sensations of Tone* [3rd ed., 1870] (1875)
10. W. Wundt, *Grundzüge der physiologischen Psychologie* (1874)
11. W. B. Carpenter, *Principles of Mental Physiology* (1874)
12. F. Galton, *English Men of Science* (1874)
13. D. Ferrier, *The Functions of the Brain* (1876)
14. G. H. Lewes, *The Physical Basis of Mind* (1877)
15. T. Ribot, *German Psychology of To-Day* [1879] (1886)
16. R. H. Lotze, *Outlines of Psychology* [1881] (1886)
17. J. H. Jackson, *Evolution and Dissolution of the Nervous System* [orig. pts, 1881–7] (1932)
18. H. Höffding, *Outlines of Psychology* [1882] (1891)
19. F. Galton, *Inquiries into Human Faculty and Its Development* (1883)
20. H. Ebbinghaus, *Memory* [1885] (1913)
21. H. Bernheim, *Suggestive Therapeutics* [1886] (1889)
22. A. Binet, *The Psychology of Reasoning* [1886] (1899)
23. E. Mach, *Contributions to the Analysis of the Sensations* [1886] (1897)
24. G. T. Ladd, *Elements of Physiological Psychology* (1887)
25. G. J. Romanes, *Mental Evolution in Man* (1888)
26. C. L. Tuckey, *Psycho-Therapeutics* (1889)

27/28. W. James, *The Principles of Psychology* (1890)

29. W. Wundt, *Lectures on Human and Animal Psychology* [2nd ed., 1892] (1894)
30. P. Janet, *The Mental State of Hystericals* [1892–4] (1901)
31. O. Külpe, *Outlines of Psychology* [1893] (1895)
32. J. M. Baldwin, *Mental Development in the Child and the Race* (1895)
33. C. L. Morgan, *Habit and Instinct* (1896)
34. E. B. Titchener, *An Outline of Psychology* (1896)
35. W. Wundt, *Outlines of Psychology* [1896] (1897)
36. J. M. Baldwin, *Social and Ethical Interpretations in Mental Development* (1897)
37. L. T. Hobhouse, *Mind in Evolution* (1901)
38. E. L. Thorndike, *An Introduction to the Theory of Mental and Social Measurements* (1904)
39. M. Prince, *The Dissociation of a Personality* (1906)
40. O. Pfungst, *Clever Hans (The Horse of Mr. Von Osten)* [1907] (1911)
41. A. Forel, *The Senses of Insects* (1908)
42. E. B. Huey, *The Psychology and Pedagogy of Reading* (1908)
43. W. McDougall, *An Introduction to Social Psychology* (1908)
44. W. D. Scott, *The Psychology of Advertising* (1908)
45. M. F. Washburn, *The Animal Mind* (1908)
46. R. Schulze, *Experimental Psychology and Pedagogy* [1909] (1912)
47. E. Claparède, *Experimental Pedagogy and the Psychology of the Child* [1910] (1911)
48. E. L. Thorndike, *Animal Intelligence* (1911)
49. H. Münsterberg, *Psychology and Industrial Efficiency* (1913)
50. E. L. Thorndike, *Educational Psychology* (1914)

Dates in square brackets indicate the original French or German editions from which these first English translations were prepared.

# An Introduction to the Theory of Mental and Social Measurements

EDWARD LEE THORNDIKE

THOEMMES PRESS

MARUZEN CO., LTD

This edition co-published in 1998 by

THOEMMES PRESS
11 Great George Street
Bristol BS1 5RR, United Kingdom

&

MARUZEN CO., LTD
3–10, Nihombashi, 2-chome, Chuo-ku
Tokyo, 103–8245 Japan

**An Introduction to the Theory of Mental and Social Measurements**
ISBN 1 85506 689 0

**Classics in Psychology, 1855–1914**
50 volumes : ISBN 1 85506 602 5

*British Library Cataloguing-in-Publication Data*
A CIP record of this title is available from the British Library

Printed in England by Antony Rowe Ltd, Chippenham

Publisher's Note

The Publisher has gone to great lengths to ensure the quality of this reprint but points out that some imperfections in the original book may be apparent.

This book is printed on acid-free paper, sewn, and cased in a durable buckram cloth.

# AN INTRODUCTION

TO THE

# THEORY OF MENTAL AND SOCIAL MEASUREMENTS

BY

EDWARD L. THORNDIKE

PROFESSOR OF EDUCATIONAL PSYCHOLOGY IN TEACHERS COLLEGE
COLUMBIA UNIVERSITY

NEW YORK
THE SCIENCE PRESS
1904

# PREFACE.

Experience has sufficiently shown that the facts of human nature can be made the material for quantitative science. The direct transfer of methods originating in the physical sciences or in commercial arithmetic to sciences dealing with the complex and variable facts of human life has, however, resulted in crude and often fallacious measurements. Moreover, it has been difficult to teach students to estimate quantitative evidence properly or to obtain and use it wisely, because the books to which one could refer them were too abstract mathematically or too specialized, and omitted altogether much of the knowledge about mental measurements most needed by the majority of university students.

It is the aim of this book to introduce students to the theory of mental measurements and to provide them with such knowledge and practice as may assist them to follow critically quantitative evidence and argument and to make their own researches exact and logical. Only the most general principles are outlined, the special methods appropriate to each of the mental sciences being better left for separate treatment. If the general problems of mental measurement are realized and the methods at hand for dealing with variable quantities are mastered, the student will find no difficulty in acquiring the special information and technique involved in the quantitative aspect of his special science. The author has had in mind the needs of students of economics, sociology and education, possibly even more than those of students of psychology, pure and simple. Indeed, a great part of the discussion is relevant to the problems of anthropometry and vital statistics. The book may with certain limitations be used as an introduction to the theory of measurement of all variable phenomena.

# CONTENTS.


| CHAPTER. | | PAGE. |
|---|---|---|
| I. | INTRODUCTION. | 1 |
| II. | Units of Measurement. | 7 |
| III. | The Measurement of an Individual. | 22 |
| IV. | The Measurement of a Group. | 41 |
| V. | The Causes of Variability and the Application of the Theory of Probability to Mental Measurements. | 61 |
| VI. | The Arithmetic of Calculating Central Tendencies and Variabilities. | 71 |
| VII. | The Transmutation of Measures by Relative Position into Terms of Units of Amount. | 85 |
| VIII. | The Measurement of Differences and of Changes. | 97 |
| IX. | The Measurement of Relationships. | 110 |
| X. | The Reliability of Measures. | 136 |
| XI. | The Use of Tables of Frequency of the Probability Surface. | 147 |
| XII. | Sources of Error in Measurements. | 157 |
| XIII. | Conclusion. References for further study. | 163 |

| APPENDIX. | | |
|---|---|---|
| I. | A Multiplication Table up to 100 × 100. | 169 |
| II. | A Table of Squares and Square Roots up to 1,000. | 190 |
| III. | Answers to Problems. Miscellaneous Problems. | 201 |

# ILLUSTRATIONS.

1–11. Illustrations of distributions of traits in individuals.
1. Memory span of B. F. A.
1A. Memory span of B. F. A.
2. Accuracy of discrimination of length of E. H.
2A. Accuracy of discrimination of length of E. H.
3. The condition of labor in the case of the Amalgamated Society of Engineers.
3A. The condition of labor in the case of the Amalgamated Society of Engineers.
4. Reaction-time of H.
5. Quickness of Movement of T.
6. Ability in addition of J. S.
7. Efficiency of perception of S.
8. Condition of a trade.
9. Attendance of a school.
10. Receipts of a sub-treasury.
11. Pulse of B.
12. Typical form of distribution.
13. Discrimination of length in 11 individuals.
14. Types of asymmetrical distribution.
15. Two distributions differing only in variability.
16. A form of distribution to illustrate the possibility of simple mathematical expression of a variable quantity.
17. A form of distribution to illustrate the possibility of simple mathematical expression of a variable quantity.
18, 19, & 20. Comparison of distributions based on few and on many records from each member of a group.
21. Distribution of height of adult men.
22. Distribution of weight of adult men.
23. Distribution of cephalic index of Alt-Bayerische skulls.
24. Distribution of length of male infants.
25. Distribution of girth of chest of adult men.
26. Distribution of strength of arm pull.
27. Distribution of body temperature at the mouth.
28. Distribution of heart rate.
29. Distribution of reaction time of college freshmen.
30. Distribution of memory span for digits.
31. Distribution of perceptive ability of 12-year-old boys.
32. Distribution of ability in controlled association.
33. Distribution of accuracy of discrimination of length.

34. Distribution of perceptive ability.
35. Distribution of ratio of attendance to enrollment in city schools.
36. Distribution of wage of cotton operatives.
37. Distribution of age of graduation from American colleges.
38. Distribution of cost per pupil of education.
39, 40. Distribution of wages of American workingmen.
41. Distribution of ratio of attendance to enrollment in city schools.
42. Distribution of income of American colleges.
43. Distribution of age at marriage of gifted men.
44. Distribution of period between marriage and divorce.
45. Distribution of size of New England families.
46. Distribution of infant mortality.
47. Distribution of age at death.
48. An irregular distribution. (Hypothetical.)
49. A regular distribution. (Hypothetical.)
50. The influence of combination of species upon the form of distribution of a group.
51, 52, 53 and 54. The influence of selection upon the form of distribution of a group.
55–67. The influence upon the form of distribution of the number and nature of causes producing the quantity.
68. Form of distribution assumed as an illustrative case of the possibility of transmuting any ordered series of known form of distribution.
69. Fig. 68 approximated by a series of rectangles.
70. Fig. 68 with base line divided to correspond to Table XXIX.
71–76. Illustrations of the inadequacy of a comparison of groups by their averages alone.
77. Comparison of boys and girls with respect to the *A* test.
78. Change in average height of 25 boys during 5 years.
79. Change in each of 25 boys in height during 5 years.
80. Change in each of 25 boys in height from 12 to 14 and from 14 to 16.
81. The relation between the refraction of air and its density.
82. The relation between lapse of time and memory.
83. The relation between two perceptive abilities.
84. The same facts as in Fig. 83, but referred to central axes.
85. The facts of Figs. 83 and 84 with the individual measures of each array replaced by their central tendency.
86. Comparison of two groups.
87. Forms of distribution.

# TABLES.

I. Variation in units of spelling ability.
II. Variation in opinions concerning units of arithmetical ability.
III. Variation in opinions concerning units of ability in controlled association of ideas.
IV. Variation in opinions concerning units of ability in controlled association of ideas.
V. Variation in opinions concerning units of ability in controlled association of ideas.
VI. Memory span of B. F. A.
VII. Accuracy of discrimination of length of E. H.
VIII. The condition of labor in the case of the Amalgamated Society of Engineers.
IX. Reaction-time of H.
X. Quickness of movement of T.
XI. Ability in addition of J. S.
XII. Efficiency of perception of S.
XIII. The condition of labor in the case of the Friendly Society of Iron Founders.
XIV. Attendance of a school.
XV. Sub-treasury daily receipts from banks.
XVI. Pulse of B.
XVII. Discrimination of length in 11 individuals.
XVIII. Two abilities differing only in variability.
XIX. Illustration of the trustworthiness of an average of a group, calculated from a few records for each member.
XX. Illustration of the trustworthiness of an average of a group, calculated from a few records for each member.
XXI. Illustration of the trustworthiness of a distribution, calculated from only a few records for each member.
XXII. Table of frequencies of the normal probability surface in terms of A. D.
XXIII. Table of frequencies of the normal probability surface in terms of $\sigma$.
XXIV. Combinations of 4 causes.
XXV. Combinations of 5 causes.
XXVI. Combinations of 6 causes.
XXVII. Combinations of 10, 15 and 20 causes.
XXVIII. Cost of education per pupil for full year's attendance in American cities.
XXIX. 100 boys ranked by their serial order with respect to a trait.

XXX. Relative frequencies of equidistant abilities of the surface of frequency of Fig. 69.

XXXI. Average distance from the average of each serial per cent. of the cases in any normal distribution.

XXXII. Average distance from the average of any continuous group in a normal distribution.

XXXIII. Abilities in the *A* test of boys and girls.

XXXIV. Growth of 25 boys from the 13th through the 17th year.

XXXV. The relation between the lapse of time and memory.

XXXVI. The relation between two perceptive abilities.

XXXVII. The relation between two perceptive abilities, each being expressed in terms of deviations from the central tendency as a zero point.

XXXVIII. The relationship of Tables XXXVI. and XXXVII. expressed in a series of measures each with the average of its related array.

XXXIX. The relationship of Tables XXXVI. and XXXVII. expressed as a series of individual ratios.

XL. Table XXXIX., with allowance made for the variability of each trait.

XLI. Calculation of $r$, using averages of arrays.

XLII. Calculation of $r$, using individual records.

XLIII. Table of values of the normal probability integral, corresponding to values of $x/\sigma$.

XLIV. Table of values of the normal probability integral, corresponding to values of $x$/A. D.

XLV. Table of values of the normal probability integral, corresponding to values of $x$/P. E.

XLVI. Table of frequencies of normal surface with Av. = 24, and $\sigma = 4$.

XLVII. Comparison of an actual distribution with the normal distribution of the same Av. and A. D.

# CHAPTER I.

## INTRODUCTION.

### *Mathematics and Measurements.*

THE power to follow abstract mathematical arguments is rare and its development in the course of school education is rarer still. For example, few of us are able to understand the symbols or processes used in the quotation on the following page. Yet it is a rather easy sample of the discussions from which the student is expected to gain insight into the theory of measurement appropriate to the variable phenomena with which the mental sciences have to deal.

It would be unfortunate if the ability to understand and use the newer methods of measurement were dependent upon the mathematical capacity and training which were required to derive and formulate them. The great majority of thinkers would then be deprived of the most efficient weapon in investigations of mental and social facts, and adequate statistical studies could be made only by the few students of psychology, sociology, economics and education who happened to be also proficient mathematicians.

There is, happily, nothing in the general principles of modern statistical theory but refined common sense, and little in the technique resulting from them that general intelligence can not readily master. A new method devised by a mathematician is likely to be expressed by him in terms intelligible only to those with mathematical training, and to be explained by him through an abstract derivation which only those with mathematical training and capacity can understand. It may, nevertheless, be possible to explain its meaning and use in common language to a common-sense thinker. With time what were the mysteries of the specialist become the property of all. To aid this process in the case of certain recent contributions to statistical theory is one of the leading aims of this book. Knowledge will be presupposed of only the elements of arithmetic and algebra. Artificial symbols will be used only when they are really convenient. Concrete illustrations will always accompany and often replace abstract laws.

Deduction of Equation of Curve of Error, from A. L. Bowley's 'Elements of Statistics,' p. 275 f.

We can now proceed to the determination of the equation of the curve of error.

The chance of $r$ successes is greatest when $r$ is the greatest integer in $pn$; this is found by the ordinary method of determining the maximum term in a binomial expansion.

Let $P$ be this maximum value $= {}^nC_{pn} \cdot p^{pn} q^{qn}$, taking the supposition for brevity that $pn$ is integral, which will not affect the proof.

$$= \frac{\underline{/n}}{\underline{/pn}\,\underline{/qn}} p^{pn} q^{qn}, \text{ for } pn + qn = n.$$

Let $P_x$ be chance of $pn + x$ white balls. Then

$$P_x = P \times \left(\frac{p}{q}\right)^x \times \frac{qn \cdot (qn-1) \cdots (qn - x + 1)}{(pn+1)(pn+2) \cdots (pn+x)}$$

$$= P \times \frac{1 \cdot \left(1 - \frac{1}{qn}\right)\left(1 - \frac{2}{qn}\right) \cdots \left(1 - \frac{x-1}{qn}\right)}{\left(1 + \frac{1}{pn}\right) \cdot \left(1 + \frac{2}{pn}\right) \cdots \left(1 + \frac{x}{pn}\right)}$$

Taking logarithms of both sides

$$\log P_x = \log P + \log\left(1 - \frac{1}{qn}\right) + \log\left(1 - \frac{2}{qn}\right) + \cdots$$

$$+ \log\left(1 - \frac{x-1}{qn}\right) - \log\left(1 + \frac{1}{pn}\right) - \log\left(1 + \frac{2}{pn}\right)$$

$$- \cdots - \log\left(1 + \frac{x-1}{pn}\right) - \log\left(1 + \frac{x}{pn}\right)$$

$$= \log P - \left(\frac{1}{qn} + \frac{1}{2} \cdot \frac{1}{(qn)^2} + \right) - \left(\frac{2}{qn} + \frac{1}{2} \cdot \left(\frac{2}{qn}\right)^2 + \right)$$

$$- \cdots - \left(\frac{x-1}{qn} + \frac{1}{2}\left(\frac{x-1}{qn}\right)^2 + \right)$$

$$- \left(\frac{1}{pn} - \frac{1}{2} \cdot \frac{1}{(pn)^2} + \right) - \left(\frac{2}{pn} - \frac{1}{2}\left(\frac{2}{pn}\right)^2 + \right) - \cdots$$

$$- \left(\frac{x}{pn} - \frac{1}{2}\left(\frac{x}{pn}\right)^2 + \right)$$

$$= \log P - \frac{1 + 2 + \cdots + (x-1)}{qn} - \frac{1^2 + 2^2 + \cdots + (x-1)^2}{2q^2n^2}$$

$$- \cdots - \frac{1 + 2 + \cdots + x}{pn} + \frac{1 + 2^2 + \cdots + x^2}{2p^2n^2} - + \cdots$$

$$= \log P - \frac{x(x-1)}{2qn} - \frac{x(x+1)}{2pn} - \frac{(x-1) \cdot x \cdot (2x-1)}{12q^2n^2}$$

$$+ \frac{x(x+1)(2x+1)}{12p^2n^2} - +$$

Let no one suppose that the foregoing statements imply that mathematical gifts and training are useless possessions for a student of quantitative mental science. On the contrary, the assumption of their absence in 'the reader' will necessitate long descriptions, roundabout arguments and awkward formulæ. If this book were written by a mathematician for the mathematically minded it would not need to be one fifth as long. If it is read by such a one, it may well seem intolerably clumsy and inelegant.

### *General Information about Measurements.*

There are, in addition to the recent studies of the general theory of mental measurements, a number of matters concerning the quantitative treatment of human nature which sufficient experience teaches thoughtful workers everywhere, but which have not been stated simply and conveniently in available form for study and reference. At present one must learn these gradually and with difficulty by himself or acquire them from the oral traditions of the laboratory or class-room. They are, for the most part, extremely simple. But that one sees them at the first glance when they are presented does not imply that he would not in nine cases out of ten fail to discover them if they were not presented. To put these at the service of all who need to know about them is the second aim of this book.

### *The Technique of Measurements.*

Although the formulæ used in expressing and comparing mental measurements are in most cases straightforward and simple, they are often so foreign to the habits acquired in connection with the arithmetic and algebra of one's school days that ready and sure use of them can be acquired only by practice. Convenient and accurate manipulation of figures is one of the many things which one learns to do by doing. A mere statement of a rule leaves one uncertain. Only after applying it a number of times does he really possess it. For example, I doubt if any one of my readers is sure that from a mere reading he understands the following, which is an accepted short method of determining the average of a number of measures: "Arrange the numbers in the order of their amount; choose any number likely to be nearest the average; add together, regarding signs, the deviations from it of all the numbers; divide this result by the number of the measures the average of which you are obtaining: add the quotient to the chosen number." To secure full mastery of every procedure taught, this book will contain many model examples and sets of problems to be worked.

*The Application of the Theory of Measurements.*

A sense of when and how to use statistical methods is even more important than knowledge of the methods themselves. The greatest benefit, therefore, will come to those who in connection with every principle established in the text, call to mind some concrete case to which the principle should be applied. The insight into the actual use of the theory of measurement thus obtained may be increased by a critical examination of the samples of quantitative studies referred to in Chapter XIII.

*The Theory of Measurements and the Special Sciences.*

This book, as the title announces, deals primarily with the theory of mental measurements. But with a few exceptions the principles and technique which it presents are applicable to all the sciences which study variable phenomena. So far, indeed, physical anthropology has been the science to take the most advantage of them, and in medicine they will perhaps find their greatest usefulness. The illustrations occasionally, and the problems frequently, come from the biological sciences. If one alters the language and replaces the illustrations from the realms of psychology and social science by similar ones from economics, vital statistics, medicine, physiology, anthropometry or biology, as the case may be, he will find the principles to hold, with an occasional obvious modification to fit the special data. The descriptions of technical procedure similarly may, after a few obvious alterations, be applied to variable measurements in general.

*The Intrinsic Interest of the Theory of Measurements.*

The author may be permitted to express his hope that those who use the book will regard its subject matter as something more than a means to the end, convenient handling of measurements. One can use ingenuity in manipulating measurements as well as in devising experiments; can use logic in working with measures as well as in working with evidence of a more impressive and dramatic sort.

Skill in expression is nowhere more required than in the task of making quantitative estimates, comparisons and relationships, brief, clear and emphatic. Statistics are, or at least may be, something beyond tabulation and book-keeping. In studying even this most elementary introduction one who is willing to use his higher intellectual powers will find something for them to do.

*The Special Problems of Mental Measurements.*

In the mental sciences as in the physical we have to measure things, differences, changes and relationships or dependencies. The psychologist thus measures the acuity of vision, the changes in it due to age, and the relationship between acuity of vision and ability to learn to spell. The economist thus measures the wealth of a community, the changes due to certain inventions and perhaps the dependence of the wealth of communities upon their tariff laws or labor laws or poor laws. Such measurements, which involve human capacities and acts, are subject to certain special difficulties, due chiefly to the absence or imperfection of units in which to measure, the lack of constancy in the facts measured and the extreme complexity of the measurements to be made.

If, for instance, one attempts to measure even so simple and mechanical a thing as the spelling ability of ten-year-old boys, one is hampered at the start by the fact that there exist no units in which to measure. One may, of course, arbitrarily make up a list of 10 or 50 or 100 words and measure ability by the number spelled correctly. But if one examines such a list, for instance the one used by Dr. J. M. Rice in his measurements of the spelling ability of 18,000 children, one is or should be at once struck by the inequality of the units. Is 'to spell *certainly* correctly' equal to 'to spell *because* correctly'? In point of fact, I find that of a group of about 120 children, 30 missed the former and only one the latter. All of Dr. Rice's results which are based on the equality of any one of his 50 words with any other of the 50 are necessarily inaccurate, as is abundantly shown by Table I. (page 8).

Economists have not yet agreed upon a system of units of measurement of consuming power. Is an adult man to be scored as twice or two and a half or three times as great a consumer as a ten-year-old boy? If an adult man's consuming power equals 1.00, what is the value of that of an adult woman?

If we measure a school boy's memory or a school system's daily attendance or a working man's daily productiveness or a family's daily expenditures, we find in any case not a single result, but a set of varying results. The force of gravity, the ratio of the weight of O to the weight of H in water, the mass of the H atom, the length of a given wire; these are, we say, constants; and though in a series of measures we get varying results, the variations are very slight and can be attributed to the process of measuring. But with human affairs not only do our measurements give varying results; the thing itself is not the same from time to time, and the individual things of a common group are not identical with each other. If we say that the mass of the O atom is 16 times the mass of the H atom, we mean that it always is that or very, very near it. But if we say that the size of the American family is 2 children, we do not mean that it is that alone; we mean that it is sometimes 0, sometimes 1, etc.

Even a very elaborate chemical analysis would need only a score or so of different substances in terms of which to describe and measure its object, but even a very simple mental trait, say arithmetical ability or superstition or respect for law, is, compared with physical things, exceedingly complex. The attraction of children to certain studies can be measured, but not with the ease with which we can measure the attraction of iron to the magnet. The rise and fall of stocks is due to law, but not to any so simple a law as explains the rise and fall of mercury in a thermometer.

The problem for a quantitative study of the mental sciences is thus to devise means of measuring things, differences, changes and relationships for which standard units of amount are often not at hand, which are variable, and so unexpressible in any case by a single figure, and which are so complex that to represent any one of them a long statement in terms of different sorts of quantities is commonly needed. This last difficulty of mental measurements is not, however, one which demands any form of statistical procedure essentially different from that used in science in general.

## CHAPTER II.

### UNITS OF MEASUREMENT.

LET us examine first a number of units that have actually been used. It is the custom to measure intellectual ability and achievement, as manifested in school studies, by marks on an arbitrary scale; for instance, from 0 to 100 or from 0 to 10. Suppose now that one boy in Latin is scored 60 and another 90. Does this mean, as it would in ordinary arithmetic, that the second boy has one and one half times as much ability or has done one and one half times as well? It may by chance in some cases, but the fact that the best one and the worst one of thirty boys may be so marked by one teacher, and during the next half year in the same study be marked 70 and 90 by the next teacher, proves that it need not. The same difference in ability may, in fact, be denoted by the step from 60 to 90 by one teacher, by the step from 40 to 95 by another, by the step of from 75 to 92 by another and even by still another by the step from 90 to 96. Obviously school marks are quite arbitrary and their use at their face value as measures is entirely unjustifiable. A 90 boy may be four times or three times or six fifths as able as an 80 boy.

It is the custom to measure the value of commodities and labor by their money price, but since a dollar in one year is evidently not necessarily equal to a dollar twenty years before, systems of index values * have been established to give a better unit. Even these index values as arranged by different statisticians differ somewhat.

For a unit of power of consumption Engel takes a child during its first year. He then calls a year-old's power of consumption 1.1; a two-year-old's, 1.2; and so on up to 3.0 for a woman 20 years or over and 3.5 for a man 25 years or over. In the United States investigation of 1890–91 the unit was taken as 100 for an adult man, 90 for an adult woman, 75 for a child 7 to 10 years old, 40 for a child 3 to 6 and 15 for a child 1 to 3. The arbitrary nature of the scale of measurement is apparent.

* The reader unlearned in economic science may neglect this illustration.

The extreme inequalities of the spelling words, treated by Dr. Rice as of equal difficulty, are shown in Table I.

TABLE I.

THE RELATIVE FREQUENCY OF MISTAKES WITHIN THE SAME GROUP OF CHILDREN FOR EACH OF 49 WORDS TAKEN BY DR. RICE TO BE OF EQUAL AMOUNT AS MEASURES OF SPELLING ABILITY.

| | By 5a Grade Girls. | By 5a Grade Boys. | | By 5a Grade Girls. | By 5a Grade Boys. |
|---|---|---|---|---|---|
| Disappoint | 24 | 13 | Frightened | 3 | 6 |
| Necessary | 23 | 19 | Baking | 3 | 6 |
| Changeable | 20 | 22 | Peace* | 3 | 6 |
| Almanac | 19 | 14 | Laughter | 3 | 6 |
| Certainly | 15 | 15 | Waiting | 2 | 8 |
| Lose | 15 | 12 | Chain | 2 | 7 |
| Slipped | 13 | 9 | Thought | 2 | 6 |
| Deceive | 13 | 7 | Weather | 2 | 4 |
| Whistling | 11 | 11 | Light | 2 | 4 |
| Purpose | 9 | 10 | Surface | 2 | 4 |
| Speech | 8 | 15 | Strange | 2 | 4 |
| Receive | 7 | 12 | Enough | 2 | 2 |
| Loose | 7 | 7 | Running | 2 | 2 |
| Listened | 6 | 9 | Distance | 1 | 6 |
| Choose | 6 | 6 | Getting | 1 | 3 |
| Queer | 6 | 5 | Better | 1 | 2 |
| Hopping | 6 | 5 | Feather | 1 | 0 |
| Believe | 5 | 8 | Rough | 0 | 5 |
| Writing | 5 | 7 | Covered | 0 | 5 |
| Smooth | 5 | 5 | Always | 0 | 4 |
| Language | 5 | 3 | Mixture | 0 | 4 |
| Neighbor | 4 | 7 | Driving | 0 | 3 |
| Learn | 4 | 2 | Because | 0 | 1 |
| Changing | 3 | 11 | Picture | 0 | 0 |
| Careful | 3 | 8 | | | |

In the three cases so far the arbitrary opinions or guesses of individuals that such and such are equal have been uncritically accepted. It is as if we should measure length in accord with some one's guess that the distance from San Francisco to Chicago equaled three times that from Chicago to New York and eight times that from New York to Boston.

The risk of accepting subjective opinion even in the cases where it is least liable to error may be illustrated by the variation in judgment, even among competent authorities (graduate students of experience in teaching), as to the relative difficulty of different parts of the following simple tests:

* Piece was scored correct.

A.

How much is $\frac{144}{9} \times \frac{27}{12} \times \frac{2}{9} \times \frac{27}{12}$?

How much is $5\frac{3}{8} + 1\frac{1}{4} - 7\frac{1}{8} + 6\frac{1}{2}$?

3. If a girl had two dollars, three five-cent pieces, two dimes and three quarter dollars, how much money would she have in all?

4. How much is $37\frac{1}{2} + 87\frac{1}{2} + \frac{250}{4} + 6 + \frac{1}{2} + 6$?

Twelve individuals assigned to examples 2, 3 and 4 the amount of credit due for successful solution of each on the basis that the successful solution of example 1 received a credit of 10. They estimated, that is, the abilities involved in doing 2, 3, and 4 in terms of the ability involved in doing 1. Their estimates varied from 8 to 20 for 2, from 5 to 20 for 3, and from 14 to 25 for 4. Their ratings in detail were (Table II.):

TABLE II.

| Example 2. | | Example 3. | | Example 4. | |
|---|---|---|---|---|---|
| Rating. | Number giving it. | Rating. | Number giving it. | Rating. | Number giving it. |
| 8 | 1 | 5 | 5 | 13 | 1 |
| 10 | 1 | 6 | 1 | 14 | 1 |
| 12 | 1 | 8 | 1 | 15 | 4 |
| 15 | 6 | 10 | 1 | 18 | 2 |
| 18 | 1 | 12 | 1 | 20 | 3 |
| 20 | 2 | 15 | 2 | 25 | 1 |
| | | 20 | 1 | | |

These variations are due to two factors; first, the variations in the opinions of the difficulty of the standard (example 1) and, second, the variations in the opinions of the difficulty of 2, 3 and 4. We may eliminate the first factor and measure the variation which would appear if the different individuals compared their opinions of 2, 3 and 4 with some objective standard by dividing their ratings for each single example by the average of their ratings for all three. When this is done their estimates still range from 6.7 to 13.7 for 2, from 3.0 to 10.9 for 3, and from 10.0 to 15.5 for 4.

So, also, if we take four individuals whose ratings were such as to show that they were practically identical in their estimates of the difficulty of 1, we find that even among just these four the ranges are 10 to 20, 5 to 15 and 15 to 25 for 2, 3 and 4 respectively.

The detailed corrected ratings were:

| | |
|---|---|
| Example 2. | 6.7, 8.2, 8.3, 9.0, 10.6, 10.9, 11.3, 12.0, 12.9, 12.9, 13.3, 13.7. |
| " 3. | 3.0, 3.8, 4.3, 4.3, 4.4, 4.5, 5.9, 6.2, 6.7, 9.0, 10.0, 10.9. |
| " 4. | 10.0, 10.9, 10.9, 11.2, 12.0, 12.4, 12.9, 12.9, 13.2, 13.3, 15.0, 15.5. |

The percentages of highest to lowest ratings of the three examples are thus 245, 363 and 155. If we choose the closest limits which will include 8 out of the 12 ratings, the percentages that the upper is of the lower limits are: for example 2, 129; for 3, 176; and for 4, 122.

B.

Write as quickly as you can besides each word in the column a word that means the opposite thing from it.

1. Vertical.
2. Ignorant.
3. Rude.
4. Simple.
5. Deceitful.
6. Stingy.
7. Permanent.
8. Over.
9. To degrade.
10. Weary.
11. To spend.
12. To reveal.
13. Genuine.
14. Level.
15. Broken.
16. Wild.
17. Part.
18. Past.
19. Permit.
20. Precise.

Eight individuals assigned to words 2 to 20 the amount of credit due for correctly writing the opposite of each of them, on the basis that the credit for writing the opposite to word 1 should be arbitrarily called 10. Their estimates varied very widely, as may be seen from the table (III.) below:

Of course, as in case A, some of this variation is due to the varying opinions, of the difficulty of thinking of the opposite of the first word *vertical*. Any one word would be an insufficient test. The influence of subjective opinion is, therefore, more fairly measured by using only those individuals whose ideas of the difficulty of

TABLE III.

| Word. | Range of Credits Given. | Detailed Credits Given. |
|---|---|---|
| Ignorant | 5–15 | 5, 6, 8, 8, 10, 11, 15, 15. |
| Rude | 6–15 | 6, 7, 7, 9, 10, 13, 14, 15. |
| Simple | 2–18 | 2, 4, 8, 9, 12, 14, 15, 18. |
| Deceitful | 2–25 | 2, 6, 8, 10, 10, 15, 18, 25. |
| Stingy | 5–20 | 5, 8, 9, 9, 10, 12, 12, 20. |
| Permanent | 9–15 | 9, 9, 10, 10, 10, 11, 14, 15. |
| Over | 1– 8 | 1, 2, 6, 6, 6, 8, 8, 8. |
| To degrade | 2–20 | 2, 5, 7, 10, 10, 14, 15, 20. |
| Weary | 6–20 | 6, 7, 9, 10, 13, 15, 18, 20. |
| To spend | 2–12 | 2, 5, 6, 8, 9, 10, 11, 12. |
| To reveal | 2–15 | 2, 8, 8, 10, 10, 11, 12, 15. |
| Genuine | 7–16 | 7, 9, 9, 11, 12, 12, 15, 16. |
| Level | 8–18 | 8, 9, 9, 10, 10, 15, 15, 18. |
| Broken | 6–18 | 6, 8, 8, 8, 8, 10, 15, 18. |
| Wild | 2–15 | 2, 6, 6, 7, 8, 8, 9, 15. |
| Part | 2–20 | 2, 5, 7, 7, 8, 9, 10, 20. |
| Past | 2–30 | 2, 6, 7, 8, 8, 9, 10, 30. |
| Permit | 8–15 | 8, 9, 9, 11, 11, 12, 15, 15. |
| Precise | 10–20 | 10, 11, 11, 14, 18, 20, 20, 20. |

the standard were alike, or by allowing for the differences. Dividing the ratings of each individual by the average of all his ratings, Table III. becomes Table IV. Table IV. contains also measures of the range in terms of the percentage that the upper is of the lower limit in the case of each word.

TABLE IV.

| Word. | Range of Credits Given. | Detailed Credits Given. | | | | | | | | Per cent. Highest is of Lowest. | Per cent. Upper is of Lower of Limits Including 5 Ratings |
|---|---|---|---|---|---|---|---|---|---|---|---|
| Ignorant | 6.6–11.6 | 6.6 | 7.0 | 7.6 | 7.9 | 9.9 | 10.7 | 11.1 | 11.6 | 176 | 147 |
| Rude | 6.9–12.5 | 6.9 | 7.9 | 8.9 | 10.0 | 10.1 | 11.4 | 12.3 | 12.5 | 181 | 125 |
| Simple | 3.6–13.6 | 3.6 | 5.3 | 8.9 | 9.6 | 10.9 | 11.1 | 11.9 | 13.6 | 378 | 134 |
| Deceitful | 3.6–15.9 | 3.6 | 7.9 | 7.9 | 11.1 | 11.4 | 12.3 | 14.0 | 15.9 | 442 | 143 |
| Stingy | 6.6–16.1 | 6.6 | 8.9 | 9.1 | 9.3 | 9.9 | 11.1 | 12.7 | 16.1 | 244 | 125 |
| Permanent | 6.4–16.1 | 6.4 | 8.5 | 11.1 | 11.4 | 11.8 | 12.3 | 13.9 | 16.1 | 252 | 125 |
| Over | 1.8– 8.9 | 1.8 | 2.6 | 3.8 | 5.9 | 6.1 | 6.2 | 7.4 | 8.9 | 495 | 151 |
| To degrade | 3.6–15.5 | 3.6 | 6.4 | 6.6 | 8.6 | 11.1 | 11.4 | 13.9 | 15.5 | 431 | 178 |
| Weary | 9.2–14.0 | 9.2 | 10.0 | 10.7 | 11.4 | 12.3 | 12.7 | 12.9 | 14.0 | 152 | 123 |
| To spend | 3.6–11.1 | 3.6 | 5.9 | 6.1 | 6.6 | 7.6 | 8.5 | 11.1 | 11.1 | 308 | 144 |
| To reveal | 3.6–14.9 | 3.6 | 7.6 | 7.6 | 8.5 | 9.9 | 10.5 | 11.1 | 14.9 | 414 | 138 |
| Genuine | 7.6–16.1 | 7.6 | 8.6 | 9.3 | 10.0 | 11.4 | 14.5 | 15.8 | 16.1 | 212 | 150 |
| Level | 9.6–17.9 | 9.6 | 9.9 | 9.9 | 10.0 | 11.6 | 11.8 | 13.6 | 17.9 | 186 | 119 |
| Broken | 7.4–14.3 | 7.4 | 7.6 | 7.9 | 8.9 | 10.5 | 11.4 | 11.5 | 14.3 | 193 | 142 |
| Wild | 3.6– 9.6 | 3.6 | 5.9 | 6.1 | 7.0 | 7.9 | 8.6 | 8.9 | 9.6 | 267 | 137 |
| Part | 3.6–12.7 | 3.6 | 6.2 | 6.9 | 7.0 | 7.6 | 8.9 | 9.2 | 12.7 | 353 | 133 |
| Past | 3.6–19.1 | 3.6 | 6.0 | 6.9 | 7.4 | 8.5 | 8.8 | 11.8 | 19.1 | 531 | 147 |
| Permit | 7.0–19.6 | 7.0 | 9.6 | 9.9 | 10.0 | 10.9 | 11.4 | 15.8 | 19.6 | 280 | 119 |
| Precise | 11.1–26.3 | 11.1 | 11.5 | 13.6 | 13.9 | 15.2 | 15.5 | 25.0 | 26.3 | 237 | 135 |

On the average the highest rating is three times the lowest, and the upper of the limits, including five ratings out of the eight, is one and three eighths times the lower.

C.

Write beside each of these words a word which means some kind of the thing that the printed words means.

1. Musician
2. Official
3. Criminal
4. Fish
5. Game
6. Study
7. Machine
8. Building
9. Furniture
10. Fruit
11. Clothes
12. Vegetable
13. Book
14. Boat
15. Tree
16. Dish
17. Plant
18. Timepiece
19. Disease
20. Pain
21. Part of speech
22. Superior officer

Seven individuals assigned to words 2 to 22, the amount of credit due, in their opinion, for correctly writing a corresponding species word for each of them, on the basis that the credit for writing a word naming a kind of musician should be called 10. In this case I will give only the estimates so corrected as to eliminate differences of opinion with respect to the difficulty of word 1, and will include, as in Table IV., the percentages of highest to lowest ratings and the percentages that the upper is to the lower of the limits that include five out of the seven ratings.

On the average the highest rating is a trifle over two and three quarters times the lowest, and the upper of the limits including five ratings is almost one and one half times the lower.

In college registration statistics the unit taken is commonly one student. The college with a score of 400 is supposed to be twice as

TABLE V.

| Word. | Range. | Detailed Credits Given. | | | | | | | Per cent. Highest is of Lowest. | Per cent. Upper is of Lower of Limits Including 5 Ratings. |
|---|---|---|---|---|---|---|---|---|---|---|
| Official | 8.1–15.9 | 8.1 | 10.0 | 12.1 | 12.1 | 14.7 | 15.5 | 15.9 | 196 | 131 |
| Criminal | 6.1–21.2 | 6.1 | 10.0 | 12.1 | 12.3 | 15.9 | 18.1 | 21.2 | 348 | 175 |
| Fish | 4.8–14.5 | 4.8 | 6.1 | 10.0 | 10.3 | 10.6 | 11.1 | 14.5 | 302 | 145 |
| Game | 4.8–13.3 | 4.8 | 8.0 | 8.1 | 9.0 | 9.1 | 10.0 | 13.3 | 277 | 125 |
| Study | 4.8–10.9 | 4.8 | 6.0 | 6.4 | 8.1 | 8.6 | 10.0 | 10.9 | 227 | 167 |
| Machine | 9.0–15.9 | 9.0 | 10.0 | 10.2 | 12.1 | 13.5 | 14.5 | 15.9 | 177 | 145 |
| Building | 6.0–11.1 | 6.0 | 7.9 | 9.0 | 9.1 | 9.7 | 10.0 | 11.1 | 185 | 123 |
| Furniture | 6.0–23.8 | 6.0 | 7.2 | 7.7 | 8.1 | 9.8 | 10.0 | 23.8 | 397 | 139 |
| Fruit | 4.5–15.9 | 4.5 | 6.0 | 7.4 | 7.7 | 8.1 | 10.0 | 15.9 | 353 | 167 |
| Clothes | 4.5–12.2 | 4.5 | 6.0 | 6.1 | 6.4 | 7.9 | 10.0 | 12.2 | 271 | 167 |
| Vegetable | 6.0–15.9 | 6.0 | 7.2 | 7.4 | 7.7 | 9.1 | 10.0 | 15.9 | 265 | 139 |
| Book | 4.8–11.1 | 4.8 | 6.0 | 6.4 | 9.1 | 10.0 | 10.2 | 11.1 | 231 | 170 |
| Boat | 3.2–18.0 | 3.2 | 4.8 | 10.0 | 10.3 | 12.3 | 13.2 | 18.0 | 563 | 180 |
| Tree | 3.2–10.0 | 3.2 | 6.1 | 7.7 | 9.1 | 9.7 | 9.8 | 10.0 | 313 | 130 |
| Dish | 3.2–10.0 | 3.2 | 7.2 | 7.4 | 7.6 | 9.0 | 9.1 | 10.0 | 313 | 126 |
| Plant | 4.8–14.5 | 4.8 | 6.0 | 8.6 | 9.0 | 10.0 | 10.2 | 14.5 | 302 | 169 |
| Timepiece | 6.0–10.0 | 6.0 | 6.4 | 7.2 | 7.4 | 7.9 | 9.1 | 10.0 | 167 | 132 |
| Disease | 7.7–15.9 | 7.7 | 9.1 | 9.8 | 10.0 | 12.1 | 13.2 | 15.9 | 206 | 145 |
| Pain | 7.9–22.7 | 7.9 | 9.0 | 10.0 | 10.9 | 12.3 | 15.2 | 22.7 | 287 | 127 |
| Part of speech | 6.0–11.6 | 6.0 | 6.4 | 7.2 | 8.6 | 10.0 | 10.2 | 11.6 | 193 | 159 |
| Superior officer | 10.0–25.9 | 10.0 | 11.1 | 16.9 | 18.1 | 19.1 | 20.3 | 25.9 | 259 | 153 |

large as the college with 200. But some students do four years' work in three, some are present only a part of the year or take only a fraction of the full course during their time of enrollment. A university with 1,000 units made up in part of teachers taking a course or two a year, of casual students that drop out to take positions and of other irregulars, might really have a smaller attendance in the true sense, a smaller influence on students, than one with only 800 units. One person equals one person as a name or physical unit, but one person studying all his time with regular and continued attendance does not equal one person taking university work as a secondary pursuit.

In measuring the fertility, or rather the reproductivity of human beings, it seems at first thought to be justifiable to use the number of children in the family as a measure. But is not the number of children who live a better measure? And may not the number of children who live through the reproductive period (say 50 years) be a still better measure, and is not perhaps the number of children, each weighted in some way by the length of his life, another measure to be considered? Surely a child who dies in five minutes is not equal as a measure of reproductivity to a child who lives sixty years. Is a child who lives only thirty years?

In the case of the 'college student' and the 'child born' we are misled by what Professor Aikins has called the 'jingle' fallacy. The words are identical and we tend to accept all the different things to which they may refer as of identical amount. A similar unthinking acceptance of verbal equality as a proof of real equality makes one measure labor on the hypothesis that any one hour is equal to any other hour of it, forgetting that the step from 7 to 8 hours *per diem* may be quite different from the step from 8 to 9 and is obviously far different from the step from 20 to 21 hours. The fallacy may be emphasized by one final illustration. Dr. Swift, in studying the effect of practice, measured motor skill by the number of time two balls could be kept tossed in the air with one hand. He took as a unit of measurement one successful pair of tosses and regarded any one such pair as equal to any other. For him, that is, the step from 0, or inability to catch and toss again at all, to 5, or the ability to catch and toss 5 times with each ball, is equal to the step from 200, or ability to keep the balls in the air 200 times without failure, to 205, or the ability to do so 205 times. But, of course, if one can do the performance 200 times he can, so far as motor skill goes, do it 205 times almost as easily, the step being nearly zero. On the other hand, the step from 0 to 5 is a very considerable gap, one which some individuals can never pass. The result of Dr. Swift's system of units is that he gets the appearance of very slow improvement in early hours of practice and very rapid improvement in late hours, a state of affairs which contradicts what is found by other investigators. Of course, 'tossing two balls once' sounds identical with 'tossing two balls once,' but it is not.

In arranging a scale of measurement one must so far as possible, (1) keep free of individual opinion, must, *i. e.*, be supported by the agreement of all qualified observers. This is most satisfactorily accomplished by so arranging observations or experiments that the trait is measured in terms of some objective units, such as seconds, millimeters, dollars. Thus, ability to memorize can be measured by time taken more justly than by amount done, for a second is a second, while one line of poetry may be easier than another line.

The accuracy of movement as tested by attempts to hit a dot can be measured more justly by actual measurement than by mere inspection; men can be ranked as to wealth better by valuations of their property than by the opinions of their neighbors. One must also (2) call equal only those things *which can be interchanged without making any difference to the issue involved.* Twelve inches can be thus interchanged with one foot without making any difference if the issue is physical measurement, but not if it is the study of language. Ten dimes can be thus interchanged with one dollar if the issue is the accounts of a store, but not if it is the area of surfaces.

Even where there are available units of amount which are commensurable they are rarely on a scale with a known zero point. Measurements of the time taken to hear a sound and react by lifting a finger are commensurable in the sense that 140 is as much faster than 150 as 150 is than 160, but an absolute zero point for slowness is not known. It is impossible, then, to argue about quickness of reaction as we can about mass or temperature.

The ability to spell correctly *disappoint* and *almanac* may be found to be equal to the ability to spell correctly *necessary* and *changeable*, but how much of an advance it is beyond the absolute zero of spelling ability can not be stated, since that absolute zero is unknown. It may be taken to be the ability to spell no word at all. But at once the objection is raised that of the many who could spell no word at all some could do so with a little training, while some would need more, and a few among the idiots could never with all possible training be gotten to spell any. In physical science we can find or infer the place where a given quantity begins, — the first increment to the absolute zero of temperature, the least quantity of mass or velocity or light, the least degree of resistance, etc.; but this is rarely our good fortune when dealing with mental facts.

The zero points from which to reckon amounts of goodness, intellect, delicacy of discrimination, memory, courage, efficiency, quickness, economic productivity, inventiveness, etc., are largely lacking. Two pounds is twice one pound not only in the sense that it takes two of the latter to replace one of the former, but also in the sense that the former represents a point on the scale of mass twice as far from the zero point as does the former. Marking 20 A's instead of

10 on a sheet of mixed capital letters, or earning $10.00 instead of $5.00, or remembering six words instead of three, or inventing four machines instead of two, can by proper choice of units be made to parallel the two-pounds-one-pound comparison in its first sense, but not in its second. For there is a less perceptive ability than that of just barely not perceiving any A; productiveness runs into minus quantities in the case of workmen who spoil raw materials with no advantageous result; there are lower grades of memory and of inventiveness than those of just not remembering one word or of just not inventing one thing.

Even when absolute zero points are not discoverable it is well worth while to consider from what point the scale we do use starts; and even when the point has to be chosen arbitrarily it is well worth while to consider the meaning and utility of different possible ones. It is the duty of the student of mental and social quantities to study the whole scale in which the units he uses lie, as well as to turn those he does use into commensurable quantities.

The influence of the zero point of a scale upon measurements made by that scale will alter the interpretation of, but not the method of making, measurements of things and conditions; but when things or conditions are compared, that is, when measurements are made of difference, change and relationships, it becomes of the utmost importance. For one of the common fallacies in the mental sciences is to compare directly the amounts of measurements made from different zero points. Another is to use arbitrarily some point along the scale as if it were an absolute zero point. Silly as it may appear, we often with mental measurements do such arithmetic as the following:

"John, who weighed 4 lbs. more than 100 lbs., has added 2 lbs. to his weight; James, who weighed 100 lbs. more than 10 lbs., has added to his weight 50 lbs. Both gained 50 per cent. and so their relative gains were equal."

"John weighs 10 lbs. more than 60 lbs. James weighs 2 lbs. more than 60 lbs. John is five times as heavy as James."

Quantities to be measured may be in a *discrete* or in a *continuous* series. A *discrete* series is one with gaps. Thus if we measure the number of children in a class we can get only integral numbers.

Sixth tenths of a man, ninety-two hundreds of a man, do not exist. There is a gap, between one man and two, two men and three, etc. A *continuous* series, such as time or velocity or intellect or wealth, is in theory capable of any degree of subdivision. Almost all mental traits and social facts due to human action are quantities in continuous series.

Any given measure of a continuous series means not a single point on the scale of measurement, but the distance along that scale between two limits. Thus if we measure the time taken to perceive and react to a signal in thousandths of a second and get .143 sec. as the measure, the .143 means commonly that that was the nearest point, that the time was nearer to .143 than to .142 or to .144; and this means, of course, that the time was between .1425 and .1435. The truer statement would be, 'A's reaction time is between .1425 and .1435.' If we measure a man's wealth in dollars as 73,448, we do not mean that he has exactly that, but that that is the nearest dollar mark. At times a measure does not mean that the individual to whom it is given is nearer to that measure than to any other on the scale used, but that he is above it and not up to the next measure. For instance, if a boy in 10 minutes gets the answers to 5 problems in arithmetic, we would commonly score him 5, but our 5 would mean, 'at least 5 and not 6.' The boy might, for instance, have almost completed the sixth in his mind, and really be, if we had a finer scale, 5.9. In mental measurements, then, any figure, say 21, may mean between 20.5 and 21.5 or between 21 and 22. It might also mean between 20 and 21, if we measured people by the point which they just did not reach, but this is almost never a useful method. The second method of measuring by the last point on the scale passed is in many mental traits the natural one and often saves labor in all sorts of measurements.*

In later operations with figures denoting measurements the method of obtaining them and their consequent meaning must be kept in mind. If a set of measures mean in each case 'from this figure to the next on the scale,' then the average calculated from them will, to represent an absolute point on the scale, need to be increased

*It is easier to put a measure between two points on the scale than to tell to which point it is nearest. Moreover, in dropping insignificant figures it is easier to drop absolutely than to add 1 place when the figure dropped is over .5 the unit of the next place.

by .5 the unit of the scale. A little experimentation and thought will create in one the useful habits of thinking of any figure for a measure on a continuous scale as representing the quantities between two limits; of realizing that for our ordinary arithmetic it represents the space from a point half-way between it and the figure below to a point half-way between it and the figure above; and of understanding that if our method of measurement makes it represent some other space, we must make proper allowance in calculation.

In many cases the measure of zero, which should mean a definite distance on the scale, either from a point below 0 to a point above it, or from 0 to the next point on the scale, means only an indefinite distance; namely, from a point above 0 to an unknown lower extreme. Thus if in measuring arithmetical ability by a test of 20 examples, we should find out of fifty boys a dozen who did none at all and should mark them zero, we could not assume that they were as a group the same distance below the 1 to 2 group as the 1 to 2 group were below the 2 to 3 group. All that would be known about the dozen boys would be that they belonged somewhere below 1. One of them might be really as far below a boy marked 1 as the latter was below a boy marked 20. In such cases we call the zero marks *undistributed* or indefinite. The same holds good, of course, for the upper as well as the lower extreme. If, in the illustration in question, a dozen boys had done all the examples perfectly and been marked 20, that score would mean, not that the boys were between 20 and 21, but that they were somewhere above 20. One should always guard against undistributed measures at the extreme of a scale.

Many mental phenomena elude altogether direct measurement in terms of amount. How many thefts equal in wickedness a murder? If the piety of John Wesley is 100, how much is the piety of St. Augustine? How much more ability as a dramatist had Shakespeare than Middleton? What per cent. must be added to the political ability of the Jewish race to make it equal to that of the Irish race? In these and similar cases the quality to be measured manifests itself objectively in so complicated and subtle effects that the task of expressing it in units of amount is hopeless.

Nevertheless, such phenomena can be measured and subjected to exact quantitative treatment. Though we cannot equate crimes, we can arrange them in a list according to their magnitude, and measure any one by its position in the list. Similarly St. Augustine, if placed in his proper rank amongst men for piety, is measured as exactly as if given a numerical score. The step from Shakespeare to Middleton in a series of dramatists ranked in order of ability is a definite measure. If a boy moves in English composition from the position of the 500th in a thousand to the position of the 74th in a thousand his gain is measured as clearly and exactly as when we measure the inches he has grown in height. Measurement by relative position in a series gives as true, and may give as exact, a means of measurement as that by units of amount.

Measurement by relative position in scientific studies is of course but an outgrowth of the common practice of mankind. The man in the street measures things not only as being so many times this, but also as being 'the biggest he ever saw' or 'about average size.'

Measures by amount of some unit have been the subject of great development in the hands of physical science, while measures by relative position have been comparatively neglected, though for the mental sciences they are of the utmost importance. The use that has been made of them already by Galton, Cattell and others gives promise that the value of a measure to which the most subtle and the most complex traits alike are amenable will in the future be more appreciated.

In measuring any person or trait by position in a series, the chief desiderata are:

1. That the arrangement of the series should not be the result of any individual's chance bias, *i. e.*, that the arrangement should represent the general tendency of a number of observers.

2. That it should not be influenced by a constant error, by bias common to all, *i. e.*, that there should be, on the whole, as much bias in any one direction as in any other.

3. That it should be on a sufficiently minute scale.

Suppose, for instance, that we wish to find the position of a certain theme among 1,000 English themes written by first-year high-school boys. No one person can, except by accident, be a perfect rater of these, for his momentary impulse or his peculiar ideals or training will overweight certain features. The combined opinion of ten equally good judges will always be truer than the opinion of any one of them. If, however, all the ten over-emphasized spelling or punctuation or humor, their combined rating would be false. Such a constant error in judgment is avoided as far as possible if judges are chosen at random.*

The value of having the themes arranged on a fine scale is; first, that the finer the scale the more precise the measure, and, second, that if a theme is then misplaced by chance it will not be displaced so far. For instance, if themes were rated simply *Good* or *Bad*, a theme near the dividing line, if put on the wrong side, would be put very far to the wrong side, *viz.*, one fourth of the total distance, whereas if they were rated in 20 divisions, one in the middle would, if put to the wrong side, be moved only one fortieth of the total distance. As a practical rule one should divide the series into as many groups as one can distinguish.

Amongst school abilities, achievements in handwriting, drawing, painting, writing English, translation, knowledge of history, geography, etc., are readily measured by serial rating, and the agreement of observers is such that great reliance can be put upon the results. In the case of more general characteristics the service of the method will be greater still, though the readiness and accuracy of the process are less.

Measures by relative position have one grave defect. Ordinary arithmetic does not apply to them. It is not possible to add '17th from top of 1,000 in wealth' to '92d from top of 1,000' as we can add fortune of $1,000,000 to fortune of $790,000. We can not say that the 10th ability from the top in 100 plus the 20th ability from the top in 100 is equal to the 14th plus the 16th. We can not equate different positions in the series with each other as we can different amounts of the same thing.

We can not, that is, on the basis of what has been so far said about measurement by relative position in a series. There are, how-

*Of course the constant errors due to the *Zeitgeist*, the general bias of the opinion of experts at any time, can be overcome only by getting ratings made fifty years apart! And it is always possible for the critic to say that the human judgments which we are invoking here, even if the best of their kind, are fallible; that the future or Deity might in perfect wisdom rate otherwise! This is true enough, but for the humble statistician the best human judgment is all that is needed. And commonly the critic's complaint that the ultimate structure of the universe contradicts a given human judgment really means that he himself does not agree with it.

ever, two possibly valid ways of transmuting a measure in terms of relative position into terms of units of amount. Given a certain condition of the series as a whole, and the statements of position can be expressed in terms of amount and made amendable to ordinary arithmetic. Given the truth of a certain theory of the amount of difference noticeable, and the same result will hold. These possibilities will be discussed in a special chapter on the measurement of mental traits by relative position.

## PROBLEMS.

1. Why would the number of men giving instruction in a university not be a fair measure of the amount of teaching done?

2. What are the faults of the following proposed as a measure of civilization: $\frac{\text{Birth-rate}}{\text{Death-rate}}$?

3. How could you get commensurate units of amount of ability in addition? In what sense could you, after obtaining such units, say that *A*'s ability in addition was twice or three times *B*'s?

4. In giving examination marks, the custom is to measure downward from a standard of perfection. Suggest a better starting point to take.

5. What are some objective units of amount used to measure criminality? What would be the advantages of measuring here by relative position?

6. Group the following measures by whole numbers, first, by using the whole numbers 14, 15, etc., to represent 13.5–14.499, 14.5–15.499, etc., and second by using 14, 15, etc., to represent 14–14.999, 15–15.999, etc.:

18.642, 17.39, 21.45, 14.81, 15.51, 17.23, 19.60, 18.42, 21.7,
15.861, 16.5, 17.92, 14.4, 19.38, 20.6, 20.5, 18.39, 17.489.

Which method would you expect to be the easier and least subject to error if one had equal amounts of practice with both? Why?

7. What is the average salary of the group represented by the following statistics?:

| | | | | | | | | | |
|---|---|---|---|---|---|---|---|---|---|
| 8 | individuals | have | salaries | above | $1,000 | and | under | $1,100 |
| 20 | " | " | " | " | 1,100 | " | " | 1,200 |
| 20 | " | " | " | " | 1,200 | " | " | 1,300 |
| 16 | " | " | " | " | 1,300 | " | " | 1,400 |
| 13 | " | " | " | " | 1,400 | " | " | 1,500 |
| 9 | " | " | " | " | 1,500 | " | " | 1,600 |
| 6 | " | " | " | " | 1,600 | " | " | 1,700 |

## CHAPTER III.

### THE MEASUREMENT OF AN INDIVIDUAL.

ANY mental trait in any individual is a variable quantity. If we measure it a number of times with a fine enough scale of measurement we get not one constant result, but many differing results. The amount of addition John Smith can do in a minute, the number of cubic feet of sand Tom Jones can dig in an hour, the food consumed by Richard Brown in a day, the weekly earnings of a particular factory — these and all facts depending on human mental traits are variable.

A constant can be measured in a single figure, but a variable for its complete measurement requires as many different figures as there are varieties of the thing. Since John Smith can add now 20, now 21, now 22, now 23 digits in a minute, his ability is not any one of these nor the average of them all, but is described truly only as 20 such and such a per cent. of the times, 21 such and such a per cent. of the times, etc. Any single figure would be but an extremely inadequate representation of his ability in addition or of that of any variable trait. The measure of a variable quantity implies a list of the different quantities appearing, with a statement of the number of times that each appeared. Such a list and statement together are called a *table of frequencies* or a *distribution* of a trait. The measure of a variable trait is thus its entire distribution or table of frequency. It is common to present a table of frequencies in a diagram in which distances along a line represent the different quantities, and the heights of columns erected along it their frequencies. Thus Figs. 1, 2 and 3 represent at once to the eye the facts given by Tables VI. to VIII. Such a figure is called a *surface of frequency*; the compound line which, with the horizontal base line, encloses it is called a *distribution curve.*

Another method of presenting graphically a table of frequencies is to draw instead of the top lines of the columns a line joining the middle points of these top lines. Figures 1A, 2A and 3A repeat Figs. 1, 2 and 3 in this form.

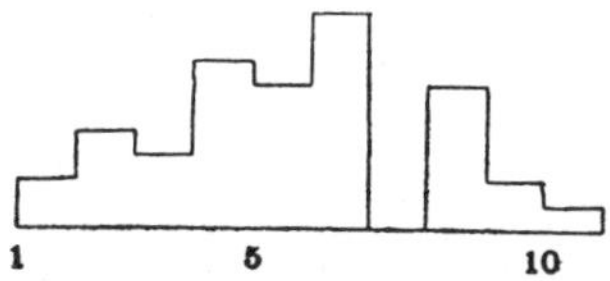


**1**

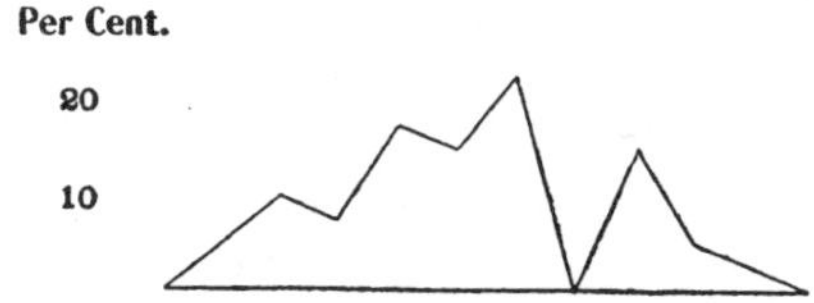


**1 A**

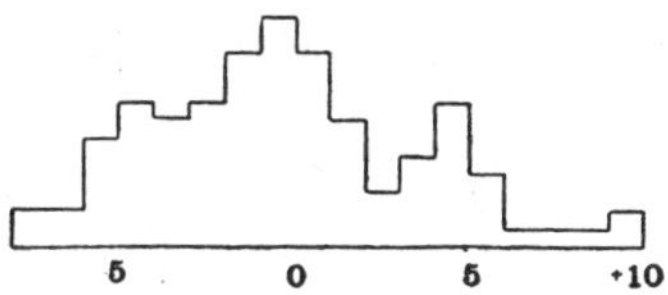


**2**

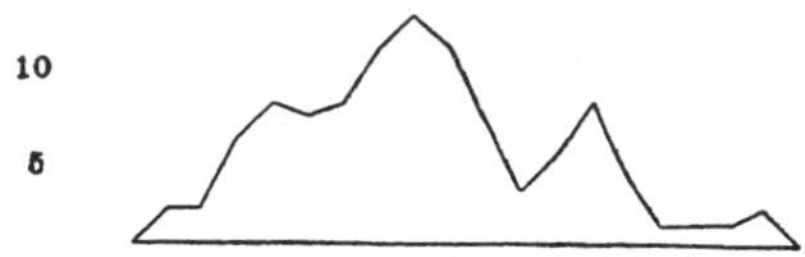


**2 A**

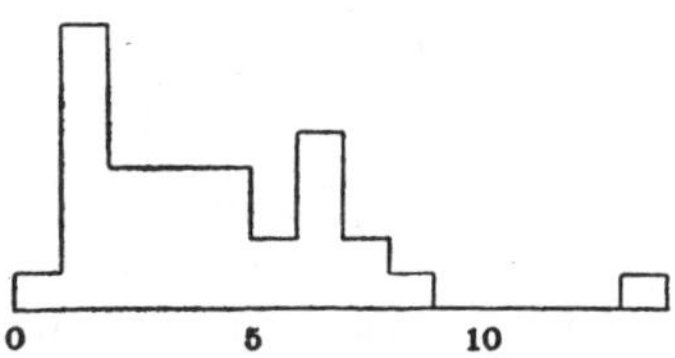


**3**

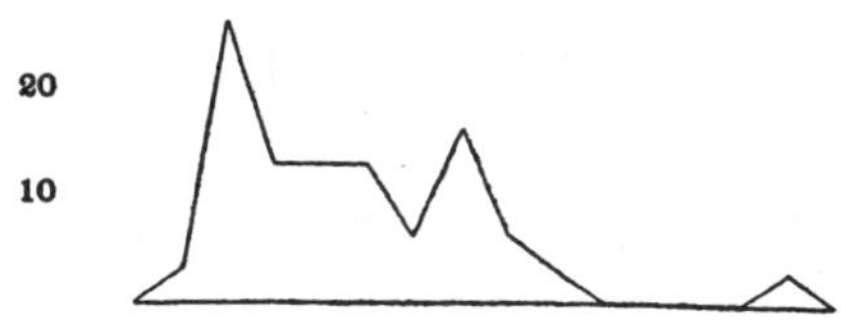


**3 A**

FIG. 1.—Surface of frequency of the ability of B. F. A. in memory span. Number of letters correctly written and correctly placed, after one hearing of a series of 12. Number of measurements = 40.

FIG. 2.—Surface of frequency of the ability of E. H. in discrimination of length. Number of millimeters error made in drawing a line to equal a 100-mm. line. Number of measurements = 100.

FIG. 3.—Surface of frequency of the opportunity for work in a trade. Number of members of the Amalgamated Society of Engineers lacking employment. Number of measurements = 31 (years).

FIG. 1A.—Same as 1, but drawn by joining mid-points of columns.

FIG. 2A.—Same as 2, but drawn by joining mid-points of columns.

FIG. 3A.—Same as 3, but drawn by joining mid-points of columns.

TABLE VI.

MEMORY SPAN OF B. F. A.

| | | | | | | | |
|---|---|---|---|---|---|---|---|
| Of a series of 12 letters | read | 1 was | correctly written | and placed | 2 times | or in | 5 % |
| " | " | 2 were | " | " | 4 | " | 10 |
| " | " | 3 " | " | " | 3 | " | 7.5 |
| " | " | 4 " | " | " | 7 | " | 17.5 |
| " | " | 5 " | " | " | 6 | " | 15 |
| " | " | 6 " | " | " | 9 | " | 22.5 |
| " | " | 7 " | " | " | 0 | " | 0 |
| " | " | 8 " | " | " | 6 | " | 15 |
| " | " | 9 " | " | " | 2 | " | 5 |
| " | " | 10 " | " | " | 1 | " | 2.5 |

There were 40 trials in all.

TABLE VII.

ACCURACY OF DISCRIMINATION OF LENGTH OF E. H.

| | | | | | | |
|---|---|---|---|---|---|---|
| In drawing a line | to equal | a 100-mm. | line an error of | — 7 mm. | occurred | 2 times. |
| " | " | " | " | — 6 | " | 2 " |
| " | " | " | " | — 5 | " | 6 " |
| " | " | " | " | — 4 | " | 8 " |
| " | " | " | " | — 3 | " | 7 " |
| " | " | " | " | — 2 | " | 8 " |
| " | " | " | " | — 1 | " | 11 " |
| " | " | " | " | 0 | " | 13 " |
| " | " | " | " | + 1 | " | 11 " |
| " | " | " | " | + 2 | " | 7 " |
| " | " | " | " | + 3 | " | 3 " |
| " | " | " | " | + 4 | " | 5 " |
| " | " | " | " | + 5 | " | 8 " |
| " | " | " | " | + 6 | " | 4 " |
| " | " | " | " | + 7 | " | 1 " |
| " | " | " | " | + 8 | " | 1 " |
| " | " | " | " | + 9 | " | 1 " |
| " | " | " | " | + 10 | " | 2 " |

There were 100 trials in all; hence per cents. = times

TABLE VIII.*

PER CENT. PER YEAR OF MEMBERS OF THE AMALGAMATED SOCIETY OF ENGINEERS IN WANT OF EMPLOYMENT DURING 31 YEARS.

| | | | | | | | |
|---|---|---|---|---|---|---|---|
| Less than 1 % | | lacked | employment in | 1 | out of 31 | years, | 3.2 % |
| 1 % to | 2 % | " | " | 8 | " | " | 25.8 |
| 2 | 3 | " | " | 4 | " | " | 12.9 |
| 3 | 4 | " | " | 4 | " | " | 12.9 |
| 4 | 5 | " | " | 4 | " | " | 12.9 |
| 5 | 6 | " | " | 2 | " | " | 6.5 |
| 6 | 7 | " | " | 5 | " | " | 16.1 |
| 7 | 8 | " | " | 2 | " | " | 6.5 |
| 8 | 9 | " | " | 1 | " | " | 3.2 |
| 9 | 10 | " | " | 0 | " | " | |
| 10 | 11 | " | " | 0 | " | " | |
| 11 | 12 | " | " | 0 | " | " | |
| 12 | 13 | " | " | 0 | " | " | |
| 13 | 14 | " | " | 1 | " | " | 3.2 |

Tables IX.–XVI., and Figures 4–11 give each the measurement of some variable trait in one individual.

* Arranged from data given by George H. Wood on pages 640–642 of Vol. 62 of the *Journal of the Royal Statistical Society*.

TABLE IX.

REACTION TIME H.

| Quantity. Thousandths of a second. | Frequency. |
|---|---|
| 120–124.9 | 9 |
| 125 | 18 |
| 130 | 35 |
| 135 | 37 |
| 140 | 43 |
| 145 | 36 |
| 150 | 38 |
| 155 | 40 |
| 160 | 38 |
| 165 | 18 |
| 170 | 24 |
| 175 | 11 |
| 180 | 15 |
| 185 | 20 |
| 190 | 10 |
| 195 | 3 |
| 200 | 4 |
| 205–209.9 | 1 |
| Total number of measures taken. | 400 |

TABLE X.

QUICKNESS OF MOVEMENT T.

| Quantity. Seconds. | Frequency. |
|---|---|
| 9.9–10.19 | 1 |
| 10.2 | 2 |
| 10.5 | 6 |
| 10.8 | 8 |
| 11.1 | 10 |
| 11.4 | 18 |
| 11.7 | 13 |
| 12.0 | 4 |
| 12.3 | 4 |
| 12.6–12.89 | 1 |
| Total number of measures taken. | 67 |

TABLE XI.

ABILITY IN ADDITION S.

| Quantity. Seconds. | Frequency. |
|---|---|
| 16–16.9 | 1 |
| 17 | 5 |
| 18 | 6 |
| 19 | 13 |
| 20 | 14 |
| 21 | 15 |
| 22 | 11 |
| 23 | 7 |
| 24 | 0 |
| 25–25.9 | 2 |
| Total number of measures taken. | 74 |

TABLE XII.

EFFICIENCY OF PERCEPTION E.

| Quantity. Number of letters seen and marked. | Frequency. |
|---|---|
| 6 | 2 |
| 7 | 5 |
| 8 | 15 |
| 9 | 20 |
| 10 | 24 |
| 11 | 16 |
| 12 | 5 |
| 13 | 0 |
| 14 | 1 |
| Total number of measures taken. | 88 |

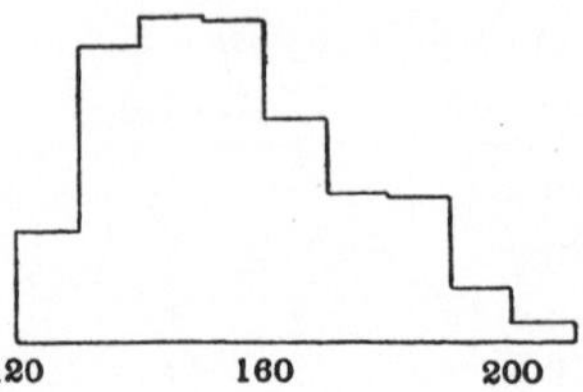


4

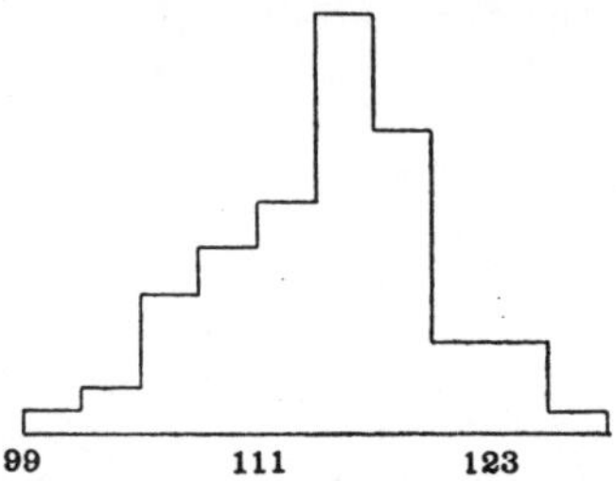


5

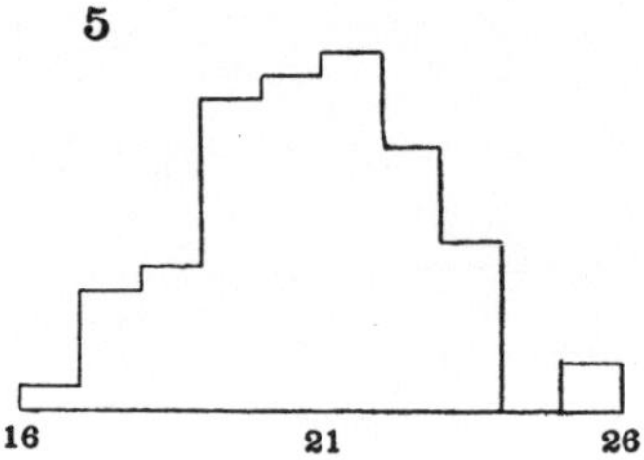


6

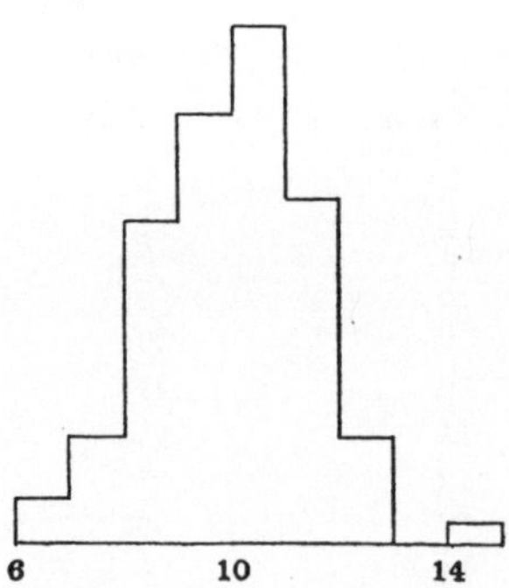


7

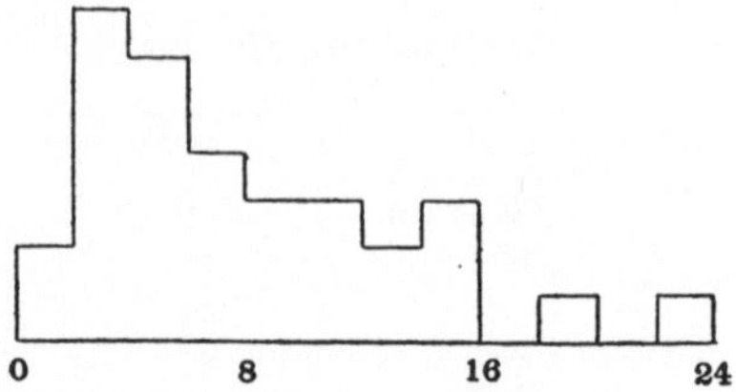


8

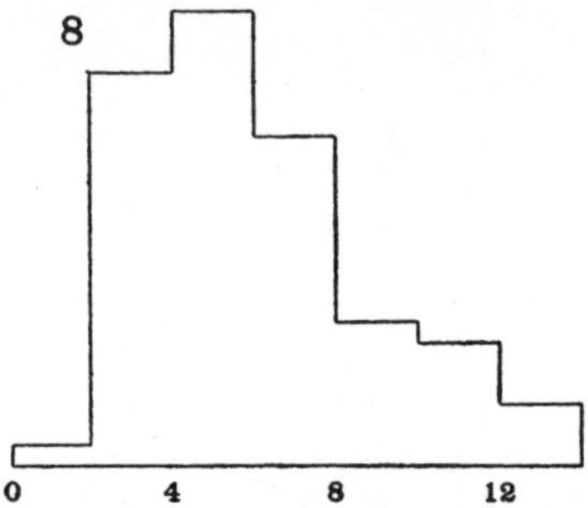


9

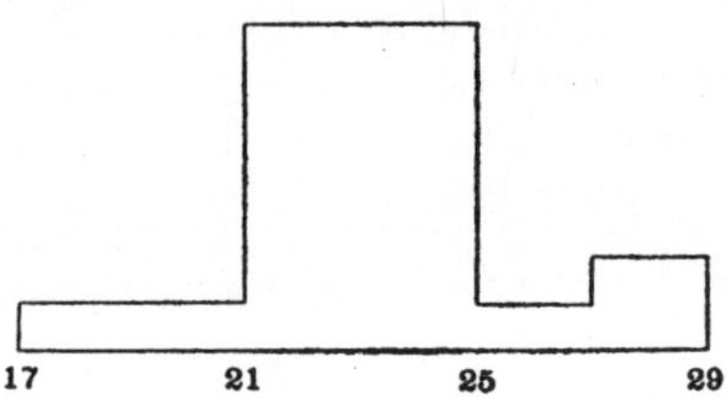


10

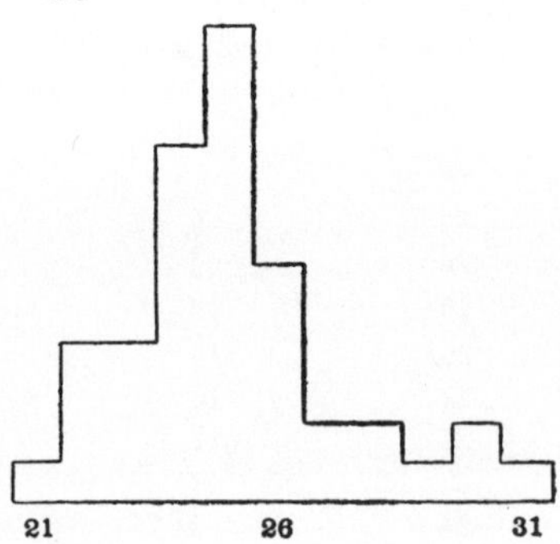


11

Figs. 4 to 11 represent the measurements of Tables IX. to XVI. in order. Fig. 4 corresponds to Table IX., Fig. 5 to Table X., etc.

TABLE XIII.

CONDITION OF A TRADE. *

| Quantity. % out of employment, by years. | Frequency. Years, |
|---|---|
| 0–1.99 | 2 |
| 2.0 | 7 |
| 4.0 | 6 |
| 6.0 | 4 |
| 8.0 | 3 |
| 10.0 | 3 |
| 12.0 | 2 |
| 14.0 | 3 |
| 16.0 | |
| 18.0 | 1 |
| 20.0 | |
| 22.0–22.99 | 1 |
| Total number of measures taken | 32 |

TABLE XIV.

ATTENDANCE OF A SCHOOL.

| Quantity. Number absent out of 139 pupils. | Frequency. Days. |
|---|---|
| 0 and 1 | 1 |
| 2 " 3 | 19 |
| 4 " 5 | 22 |
| 6 " 7 | 16 |
| 8 " 9 | 7 |
| 10 " 11 | 6 |
| 12 " 13 | 3 |
| | 74 |

TABLE XV.

DAILY EXCHANGES OF A CLEARING HOUSE.

| Quantity. $10,000,000 s. | Frequency. |
|---|---|
| 17 to 19 | 1 |
| 19 " 21 | 1 |
| 21 " 23 | 7 |
| 23 " 25 | 7 |
| 25 " 27 | 1 |
| 27 " 29 | 2 |
| Total nnmber of measures taken | 19 |

TABLE VIX.

PULSE OF B.

| Quantity. Time taken for 30 beats. In seconds. | Frequency. |
|---|---|
| 21 | 1 |
| 22 | 4 |
| 23 | 4 |
| 24 | 9 |
| 25 | 12 |
| 26 | 6 |
| 27 | 2 |
| 28 | 2 |
| 29 | 1 |
| 30 | 2 |
| 31 | 1 |
| | 44 |

If it were necessary to pick some one kind of distribution as the best representative of all these, one would choose that approached by Figs. 1, 2, 5, 6, 7. In them we see the separate measures distributed symmetrically about a single central measure, and decreasing in frequency as we pass from the central measure toward either extreme, slowly at first, then more rapidly and then more slowly.

* Friendly Society of Iron-founders' report, arranged from data given by G. H. Wood, Journal of the Royal Statistical Society, Vol. 62, pp. 640–642.

They follow roughly the type shown in Fig. 12. But obviously there is no one kind that adequately represents all. The number of central types need not be one, and the variations from the central type may occur in all sorts of ways.

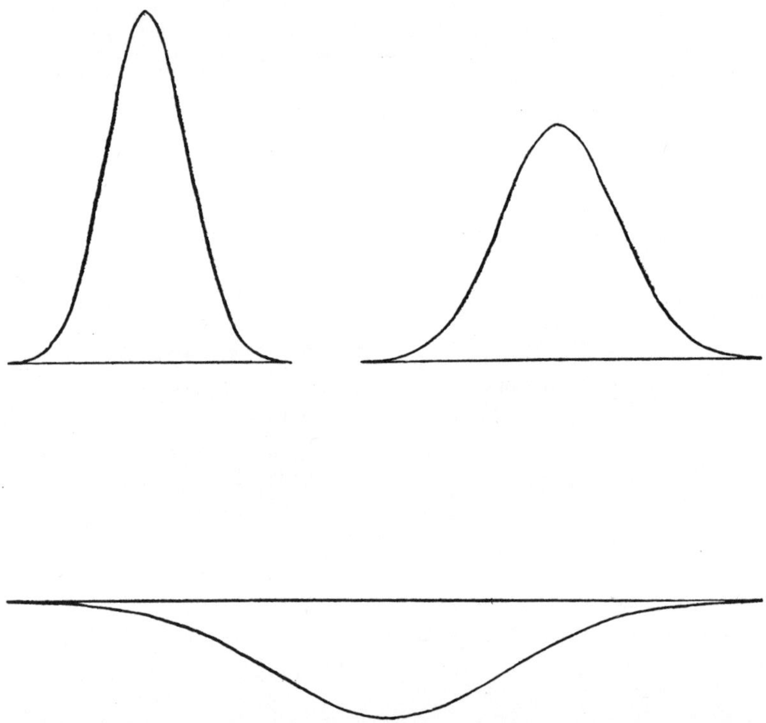

FIG. 12.—Type of distribution to which variable traits in individuals often roughly approximate. The three forms represent the same type of distribution, the only difference being in the variability.

Indeed, even in the same trait there may occur among different individuals different types of distribution. Table XVII. and Fig. 13 illustrate this in the case of the accuracy of a certain kind of perceptive process in eleven individuals. The individuals were chosen at random and so give an impartial representation of the fact.

## TABLE XVII.

### Drawing a Line Equal to a 100-mm. Line Seen.

| Quantity Length. | Frequencies in the case of 11 individuals. | | | | | | | | | | |
|---|---|---|---|---|---|---|---|---|---|---|---|
| | $M$ | $S$ | $F_1$ | $C$ | $L_1$ | $R_1$ | $H$ | $F_2$ | $L_2$ | $R_2$ | $B$ |
| 81 | | | | | | 1 | | | | | |
| 2 | | | | | | | | | | | |
| 3 | | | | | | 1 | | | | | |
| 4 | | | | | | | | | | | |
| 5 | | | | | | 4 | | | | | |
| 6 | | | | | | 2 | | | | | |
| 7 | | | | | | 3 | | | | | |
| 8 | | | | | | 1 | | | | | |
| 9 | | | | | | 1 | | | | | |
| 90 | | | | 1 | | 2 | | | | | |
| 1 | | | | | | | | | 1 | | |
| 2 | 1 | 1 | | | | 2 | | | 1 | | |
| 3 | | 3 | | 1 | | 1 | 2 | | 1 | | 1 |
| 4 | | 4 | | | | 2 | 2 | | 1 | | |
| 5 | 1 | 6 | | 1 | | 3 | 6 | | 1 | 11 | |
| 6 | | 9 | | 7 | | 3 | 8 | | 3 | 7 | |
| 7 | 0.5 | 8 | | 6 | | 2 | 7 | | 2 | 4 | 3 |
| 8 | 2 | 8 | 4 | 16 | 12 | 2 | 8 | | 4 | 5 | 2 |
| 9 | 2 | 12 | 3 | 9 | 39 | 6 | 11 | 7 | 4 | 3 | 3 |
| 100 | 2 | 12 | 8 | 23 | 38 | 6 | 13 | 25 | 4 | 16 | 9 |
| 1 | 9 | 14 | 8 | 9 | 9 | 4 | 11 | 31 | 7 | 8 | 1 |
| 2 | 7 | 5 | 9 | 9 | 2 | 9 | 7 | 24 | 10 | 11 | 12 |
| 3 | 5 | 8 | 13 | 12 | | 7 | 3 | 11 | 12 | 8 | 9 |
| 4 | 9 | 3 | 13 | 1 | | 8 | 5 | 2 | 13 | 3 | 12 |
| 5 | 8.5 | 7 | 9 | 2 | | 6 | 8 | | 9 | 7 | 18 |
| 6 | 7.5 | | 6 | 2 | | 6 | 4 | | 14 | 6 | 12 |
| 7 | 10.5 | | 5 | | | 4 | 1 | | 7 | 5 | 4 |
| 8 | 8.5 | | 2 | | | 5 | 1 | | 2 | 0 | 5 |
| 9 | 5 | | 1 | 1 | | 5 | 1 | | 1 | 2 | 4 |
| 110 | 5 | | 2 | | | 3 | 2 | | 4 | 4 | 3 |
| 1 | 4 | | 1 | | | | | | 0 | | |
| 2 | 5.5 | | 3 | | | 1 | | | 1 | | 1 |
| 3 | 3.5 | | 2 | | | | | | | | 1 |
| 4 | 2.5 | | 3 | | | | | | | | |
| 5 | 0.5 | | 3 | | | | | | | | |
| 6 | 0 | | 0 | | | | | | | | |
| 7 | 0.5 | | 2 | | | | | | | | |
| 8 | 0.5 | | 0 | | | | | | | | |
| 9 | | | 1 | | | | | | | | |
| 120 | | | 0 | | | | | | | | |
| 1 | | | 0 | | | | | | | | |
| 2 | | | 1 | | | | | | | | |
| 3 | | | 0 | | | | | | | | |
| 4 | | | 1 | | | | | | | | |

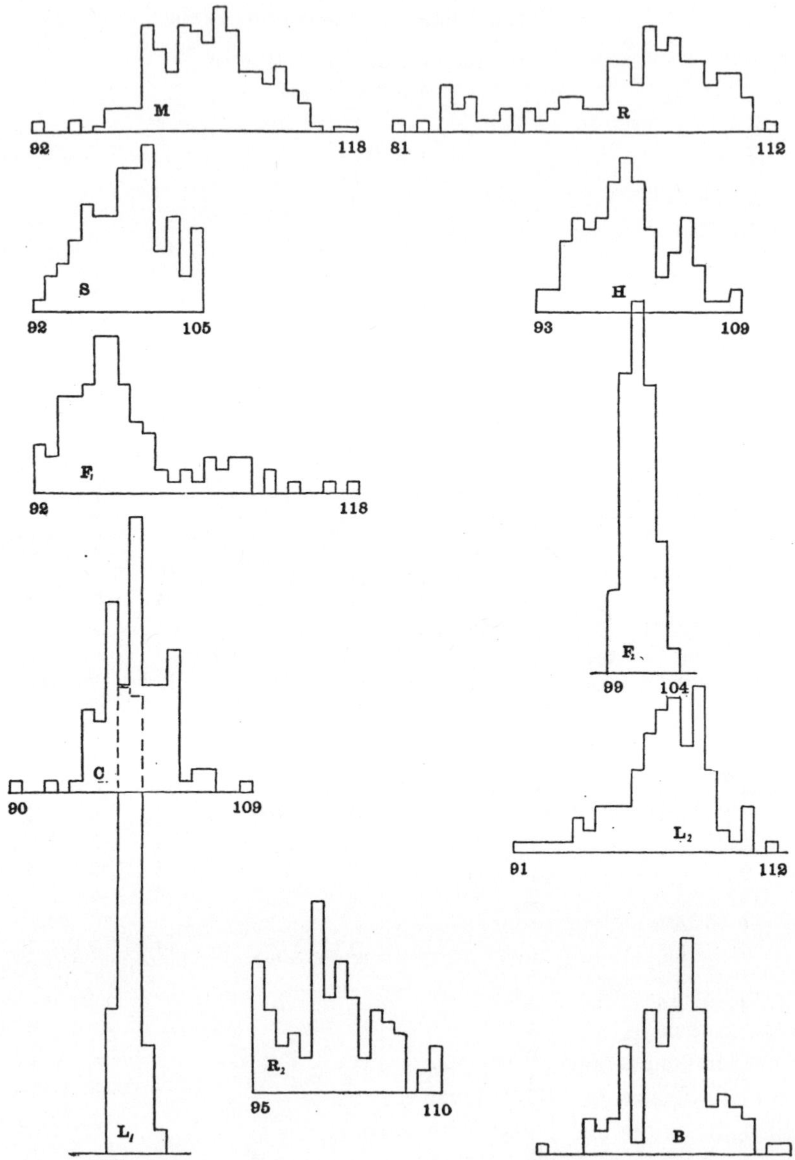


FIG. 13.—The surfaces of frequency that correspond to the tables of frequency of Table XVII.

Before discussing further the treatment of a measure expressed in a table of frequencies, it will be well to examine some clearer cases of a hypothetical nature. Suppose, for example, that measures were at hand: (1) of the daily consumption of wealth by an individual, (2*a*) of the hours worked daily by an earnest laborer, whose union did not permit more than an eight-hour day, (2*b*) of the rate of adding of a practiced accountant, (3*a*) of the amount of alcohol imbibed daily by a dipsomaniac, and (3*b*) of daily arrests for drunkenness in a city.

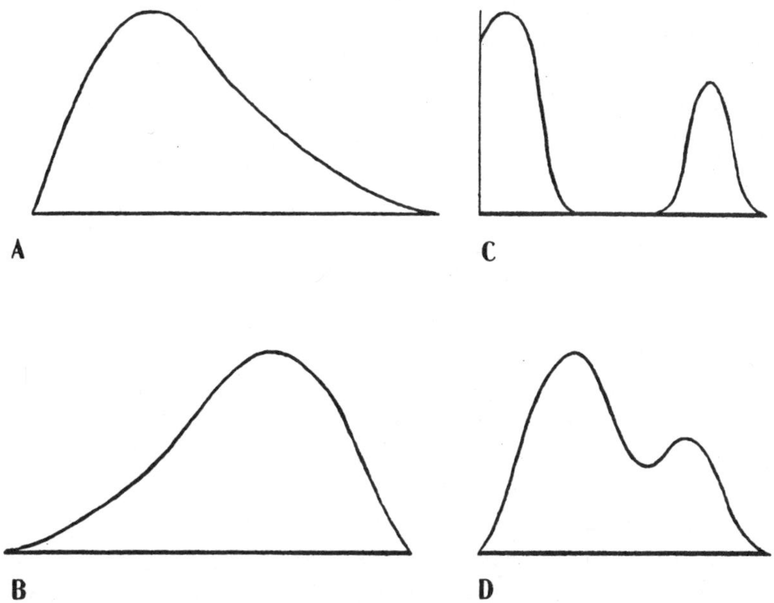


FIG. 14.

An individual who most frequently consumes two dollars' worth in food eaten, clothes worn out, minor luxuries, etc., may consume five dollars' worth by an expensive dinner, ten dollars' worth by burning up his coat, or a hundred dollars' worth by breaking a vase or overdriving a horse. He can not consume less than zero. The range of distribution limited below, runs out above a long way for practically every one. Its form will be that of Type A in Fig. 14, a form skewed toward the high end.

The laborer can not work over eight hours, but will less and less readily suffer a greater and greater decrease from that amount due to weather, employer's convenience, etc. The frequency of seven-hour days will be much below that of eight; that of six-hour days below that of seven, etc. I omit from consideration Sundays and holidays. The form of distribution will be that of Type B in Fig. 14, being skewed toward the low end. So also the practiced accountant will work in most cases near his best rate; but while nothing can raise him far above his customary rate, distraction of attention by outside stimuli, fatigue or bewilderment may drag him far below it.

The periodic dipsomaniac drinks either a great deal or little or none, according to the presence or absence of the fit of craving. The distribution of the daily amount of liquor drunk by him will therefore have two points of great frequency, with very slight frequencies for intermediate points, as shown in Type C of Fig. 14. The city's daily arrests for drunkenness will show a similar, though not so pronounced, composition of great numbers due to Saturdays, Sundays and holidays, and smaller numbers due to ordinary days. See Type D in Figure 14.

These hypothetical cases emphasize types of clear departure from the common bell-shaped form, and illustrate the insecurity of any answer to our next question, *viz.*, *How can the main meaning of a table of frequencies be expressed in one or two single figures capable of treatment by ordinary arithmetic, or in some simple algebraic equation?*

It is customary to use for any trait in an individual his average measure, but obviously, though the averages of A and B in Table XVIII. and Fig. 15 are identical, their abilities are widely different, A being a very constant performer, while B is the reverse. Again the average of the man's daily consumption of wealth figured in 14A not only does not distinguish him from some one less given to extreme prodigality who in general lives on a higher material plane, but also gives no idea of his common daily expenses. So also the average performance of the accountant does not tell what is really desired, namely, what the man can do under proper conditions. With a case like that of the dipsomaniac the average grossly misrepresents the facts to all readers who follow the common habit of expecting an average to approximate to the individual's typical performance. An average is mathematically only the sum of a set of measures divided by their number. It represents the typical measure of the set only when there is but one typical measure and when the set of measures are symmetrically disposed about it. There may be and often is more than one type of measure prominent, and the distribution may be and often is skewed instead of symmetrical.

It is clear that in every case there are needed at least two measures, one of the general tendency or typical performance, or measure

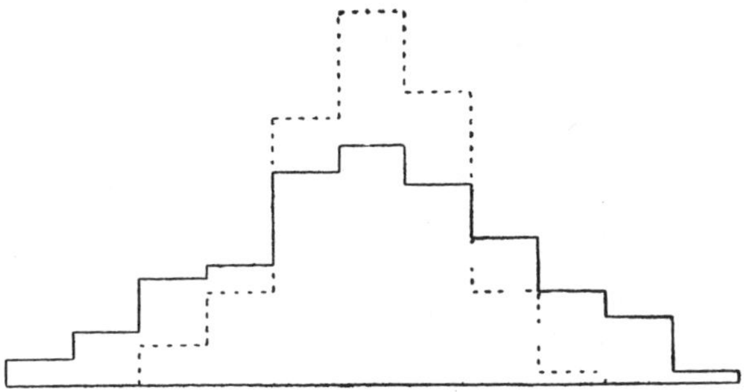

FIG. 15. — The dotted line gives A's ability, the continuous line gives B's. (This imaginary case is paralleled by many real instances. See, for instance, C and $L_1$ in Fig. 13.)

about which the individual measures cluster, the other of the variability or deviations from the type or closeness of the clustering. If there are two or more distinct tendencies or types of performance for an individual a measure for each is needed. If the deviations from

TABLE XVIII.

| Quantity. | Frequency. For A. | Frequency. For B. |
|---|---|---|
| 21 | | 2 |
| 22 | | 4 |
| 23 | 3 | 8 |
| 24 | 7 | 9 |
| 25 | 20 | 16 |
| 26 | 28 | 18 |
| 27 | 22 | 15 |
| 28 | 7 | 11 |
| 29 | 1 | 7 |
| 30 | | 5 |
| 31 | | 1 |

the type follow different gradations above and below it, as in skewed distributions, separate measures are needed for those above and below. In general, so far as the frequencies of different degrees of deviation follow no simple law, no single figure can describe them.

It is customary to use for a measure of any mental trait's variability in an individual the average or mean deviation of the separate

3

measures from their average. But the considerations just mentioned and the fact that variability may be extremely irregular disallow any such naïve procedure. The amount drunk by the dipsomaniac in the illustration really varies little, provided we take him in drinking fits alone or in sober conditions alone, but the single figure of the mean variation would picture a man of wide range day by day.

The only set of figures which adequately represent a variable measure in an individual are those from which the entire table of

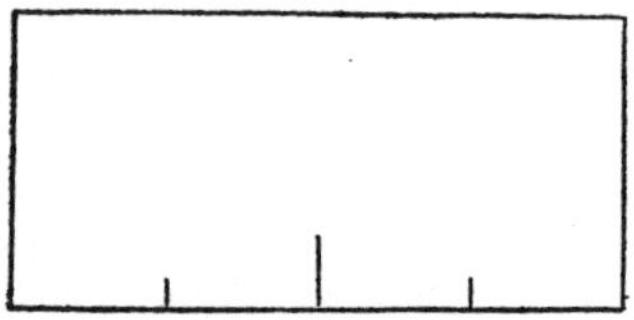

Fig. 16.

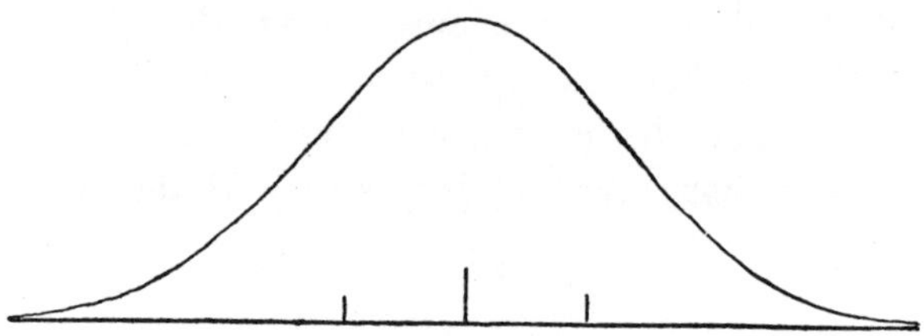

Fig. 17.

FIG. 16. — Distribution of a quantity with Average 10; Average Deviation from it 2; form of the surface of frequency, a rectangle.

FIG. 17. — Distribution of a quantity with Average 10; Average Deviation from it 2; form of the surface of frequency, that of the normal probability integral.

frequency can be calculated, which present it in briefer space and more convenient manner, but unaltered. In certain cases two or three figures with a statement of the general form of the distribution could do this. Thus, "Av. 10. Average deviation 2. Form of distribution, a rectangle," tells us that the distribution is that of Fig. 16. So also "Av. 10. Average deviation 2. Form of distribution, that of the surface of frequency of the normal probability integral," tells the student who is acquainted with certain facts that the distribution is that of Fig. 17.

It is obvious that if the distribution does not take some regular form it can not be represented by a simple algebraic expression.* In certain cases, where it does take such a regular form, it can be so represented. Thus if a man's earnings ranged from $A$ to $B$ per day and were one as often as another of these values, the surface of frequency would be the rectangle with base $AB$, and with height determined by the number of individual measures and the scale taken for the frequencies. In algebraic language, letting $x$ equal the quantity and $y$ the frequency, $y = K$ or 0, $K$ for values of $x$ between $A$ and $B$, 0 for all other values of $x$.

If a man's daily earnings varied from $A$ to $B$, decreasing in frequency in arithmetical progression as the amount increased until at $B$ the frequency was 0, the surface of frequency would be made up of such a series of rectangles of equal base as would be inscribed in a right-angled triangle. The rate of decrease would decide the slope of the triangle's hypotheneuse. As the amount of earnings was distributed on a finer and finer scale the surface of frequency would more and more approach a right-angled triangle, the mode being one side. $Y$ would equal $K(B - x)$ within the limits of $x = A$ and $x = B$, and 0 for all other values of $x$. $K$ would be a constant measuring the rate of decrease.

If the man's earnings varied from $A$ to $B$, the frequency increasing in arithmetical progression from 0 at $A$ up to $C$ and decreasing regularly in the same progression from then on to 0 at $B$, the surface of frequency would approach as a limit, a finer and finer scale of amounts being used, an isosceles triangle with base $AB$. The slope of its two sides would be decided by the rate of increase and decrease as measured by a constant $K$. $Y$ would equal $K(x - A)$ for values of $x$ from $A$ to $C$, $K(B - x)$ for values of $x$ from $C$ to $B$, and 0 for all other values of $x$.

If a man's earnings on any one day were due to the action of one combination out of all the possible combinations, all equally likely to occur, of an infinite number of causes equal in amount and independ-

* As the scale of measurement is made finer the top of the surface will of course tend to become a continuous line. For it then some mathematical expression can be discovered. The relation of the vertical distance representing frequency to the horizontal distance representing quantity is, of course, the relation actually shown in the curve and to be shown algebraically. The frequency is commonly called $y$ and the quantity $x$. Or if the distribution curve is drawn in the manner shown on page 23 (by joining the middle points of the top of the rectangles), the inquiry may be made as to the expression which will best satisfy that series of points.

ent of each other, the distribution would be of the sort shown in Figures 12 and 17. The equation would be (if $P =$ the maximum ordinate)

$$y = Pe^{\frac{-x^2}{2npq}} \quad \text{or} \quad y = e^{-x^2}$$

or some specialized form (*e.g.*,

$$y = \frac{1}{\mu\sqrt{2\pi}} e^{\frac{-x^2}{2\mu^2}},$$

in which case $\mu$ gives a measure of the variability of the trait). *

This last case is identical with the last case of the description of a distribution by two single figures. The surface of frequency thus obtained is that to which the bell-shaped distributions often approximate. If it is constructed from an infinite number of individual measures, its average, mode and median exactly coincide. They are approximately coincident when the distribution is of only a small number of measures, the differences between them being in the long run greater the smaller the number of measures is. A deviation of any amount above the average is with an infinite number of measures of the same frequency as a deviation of the same amount below. It is of approximately the same frequency when a limited number of measures are taken. The frequency of deviations decreases with their amount, first slowly, then rapidly and then slowly again. It is called the curve of error or the normal type of distribution. Its properties will be more fully described in Chapters IV. and V. The frequency with which traits in an individual are approximately so distributed, the nature of the traits in such cases and the closeness of the approximation, have hardly been studied.

Concerning the algebraic expression of a table of frequencies, the warning of page 34 must be repeated:

The only equation or set of equations which adequately represent a variable measure in an individual are those from which the entire table of frequencies can be calculated, which present it in briefer space or more convenient form, but unaltered.

From all these considerations a few simple rules emerge:

1. The real measure of a variable trait in an individual is the table of frequencies.

* The reader unfamiliar with higher algebra will have to take this on faith.

2. Beware of inferring too much from any single measure or few measures of an individual. *

3. Always turn a series of measures into a table of frequencies before inferring anything from them.

4. Never replace a table of frequencies by mere measures of their average and mean variation until simplification is necessary.

5. Never write about an average or a mean variation without an accurate description of the type of distribution whence it came. It is probably wise to print every distribution in detail.

When the distribution can be described by two measures, one of general tendency and one of variability, and when it is necessary to use such measures even though they give only an inaccurate description, the following points should be borne in mind:

Two other measures of a variable trait, the median and the mode, are often more serviceable than the average and are commonly useful in addition to it.

The median is the measure above which and below which are equal numbers of the separate measures.

The mode is the most frequent measure.

The mode is especially helpful in the case of distributions showing two or more types of performance by the same individual, for each type can be represented by a different mode and its relative importance by its mode's frequency.

The following characteristics of the different measures may help to decide which is the best to use in any given case:

The mode is the most easily and quickly determined. It is not so reliable a measure as the others. That is, the actual mode obtained from a given number of cases will not be so near the true mode as will the actual average to the true average. In reality, however, since the mode is commonly taken on a much rougher scale than the average, it is really often just as reliable, only less precise. It is hardly at all influenced by extreme measures or erroneous measures. It is entirely unambiguous and does not mislead a reader into thinking that all the individual measures of a group are very closely near it.

The median is more easily determined than the average. It is not so precise as the average, is very little influenced by extreme or erroneous measurements and is unambiguous.

* The number needed will be discussed in Chapter X.

The average is determined only with considerable arithmetical work, but this same work gives the variability as well. It is more precise than the mode or the median because the amount of every measure plays a part in determining it, but for this very reason it is more influenced by extreme or erroneous measures. The average is the measure in common use and has the advantage of being a familiar term, and at the same time the disadvantage of leading untrained readers to think that the abilities of which it is the average are closely clustered about it.

Measures of the variability or closeness of clustering of the individual measures are of two sorts. There are measures of the average of the deviations of the individual measures from their central measure, and measurements of the limits above and below the central measure which include a certain proportion of all the individual measures.

Of the first sort we have the average deviation, which equals the average of the deviations of the individual measures from their average, median or mode; and the mean square deviation or standard deviation, which equals the square root of the average of the squares of the deviations of the individual measures from their average, median or mode. Of the second sort the measure in common use is the probable error, or P. E., which gives the distance which must be taken above and below the average, median or mode, in order to include between the two limits thus obtained 50 per cent. of all the individual measures. We can, however, calculate in a similar way the limits needed to include 10, 20, 75, 90 or any other per cent. of the individual measures, and can reckon deviations from any point as well as from the central tendency, if we choose.

Strictly speaking, measures of the first sort are calculated only from the average, but it is entirely allowable to reckon them from the mode or median if a statement is made that this is done.

Measures of the first class are the more reliable in the sense that if the measures for the separate trials are reliable the same number gives an average deviation or deviation of mean square more exactly than it gives the probable error. They are, however, more influenced by erroneous or extreme measures.

In the case of skew distributions the mode is in general the most advantageous measure of general tendency; the variabilities above and below it should be given separately.

In the case of multimodal distributions the different modes should each be stated; the total table of frequencies should be analyzed into different distributions, one for each of the different modes; these distributions should be treated separately by the above rules.

The statement of the limits needed to include 20 to 30 per cent. of the cases is often a convenient expression of typical performance, giving, as it does, a wide mode.

If the measures of an individual are not in terms of amount, but are simply a ranked series of acts of kindness, or poems, or crimes, or examination papers in Latin or geography or English themes, the only measures of central ability that we can use are, of course, the mode or the median; of these the mode is commonly the most instructive. The only measures of variability that can be used are measures by limits including a given percentage.

Finally, it is a safe rule to ask concerning any figure derived from a distribution of a variable trait, 'Just what real quantity in the man does this figure represent?' and to use the figure only when a definite answer can be given.

## Problems.

8. Express in tables of frequencies and surfaces of frequency the following facts:

Ar., being measured with respect to his memory span for letters 40 times, showed the following abilities, in terms of the number of words remembered in their correct positions: 7, 6, 7, 5, 8, 2, 10, 6, 7, 8, 3, 8, 6, 9, 6, 10, 6, 8, 6, 4, 9, 6, 10, 8, 6, 8, 5, 6, 4, 8, 10, 7, 4, 7, 6, 9, 1, 11, 7, 7.

D., being measured in the same trait 40 times, showed records of: 5, 4, 1, 6, 5, 5, 8, 4, 6, 5, 5, 5, 4, 6, 4, 4, 5, 7, 2, 5, 5, 4, 5, 4, 6, 9, 4, 3, 0, 5, 5, 6, 5, 6, 3, 8, 4, 5, 5, 3.

9. Which is the more variable, Ar. or D.?

10. What is the average deviation of each from his mode?

11. In which case is it almost a matter of indifference whether the general tendency is expressed by the average or by the median or by the mode?

12. Is it a matter of indifference in the case given in question 13?

13. The percentages of workmen out of employment in England were, for different years from 1860 to 1891, 0–.99, no years; 1–1.99, 9 years; 2–2.99, 10 years; 3–3.99, 4 years; 4–4.99, 7 years; 5–5.99, 1 year. Comparing this table of frequencies with that given in Table XIII., which was the worse on the whole, the condition with respect to getting employment of workmen in general or that of the members of the Friendly Society of Iron Founders? Which was the more variable?

14. Why would not the average be a sufficient measure of the general tendency of an individual's body-temperature?

15. What would be the probable form of distribution of the daily traffic of a city's street-railroad system?

# CHAPTER IV.

## THE MEASUREMENT OF A GROUP.

THE sciences of human nature commonly use measures of individuals only in order to get measures of groups. Not John Smith's spelling ability, but that of all fifth grade boys taught by a certain method; not A's delicacy of discrimination of weight, but that of all men; not B's wage, but that of all railroad engineers during a certain period; not the number of C's children, but the productivity of the English race as a whole; not individuals, but groups, are commonly to be measured, compared and argued about.

The customary expression of a trait or ability in a group is its average, and the use of an average here, as before, points to the variability of the fact. We do not seek the average law of gravity, or the average ratio of amount of oxygen to amount of hydrogen in an atom of water, or the average velocity of sound. It is because of the unlikeness, the variability, of even the most similar human individuals in even the most constant human qualities that we are forced to use averages at all.

An average no more represents the different abilities of the members of a group than it did the different measures of a trait in a single individual. The thing, trait A in group X, is a variable quantity and is measured only by a list of the different degrees of the trait found in all the individuals of the group, with a statement of the number of times each appears. A table of frequencies or surface of frequency will be the adequate measure here, as before. The measure of a trait in a group is its total distribution, and this total distribution is simply all the separate measures of the individuals making up the group.

The measure taken for each individual may be his average or his most frequent ability or highest ability shown, or lowest ability shown, or ability exceeded in 50 per cent. of his trials, or ability exceeded in 70 per cent. of his trials, or variability, or total distribution, or any other characteristic of "individual in group X."

Most frequently some measure of central tendency is the one to be used. In such cases the individual measures may be from very few trials without doing much harm. In fact, an accurate representation of the ability of a group may arise from very inaccurate measurements of the individuals in it; for instance, from measurements from only a single record from each individual. The reason

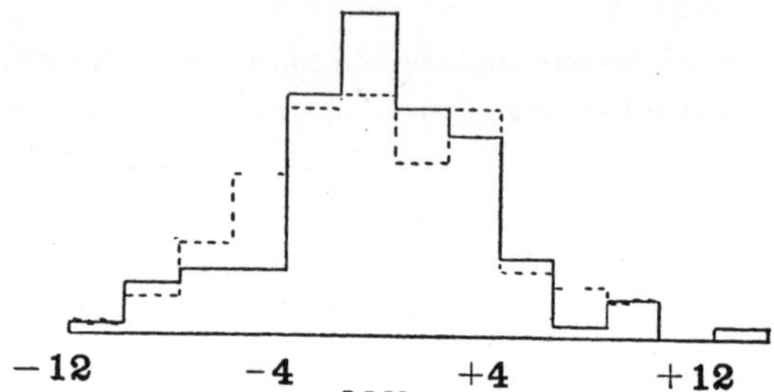


18

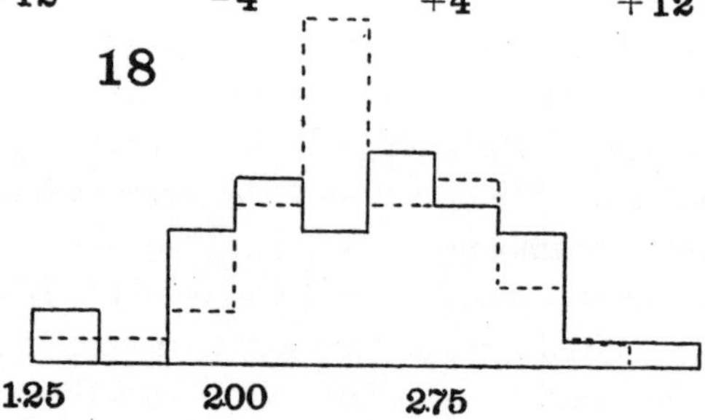


19

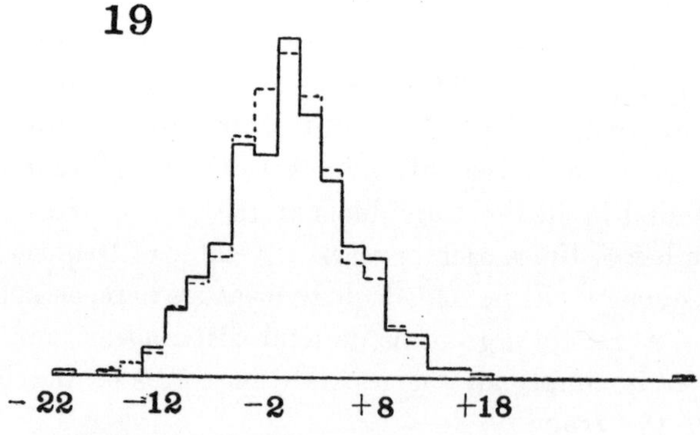


20

FIGS. 18, 19 and 20 present graphically the facts of Tables XIX., XX. and XXI. respectively.

is, of course, that the errors being chance errors, the too high rating of *A* is counterbalanced by the too low rating of *B*, and so on, so that with hundreds of cases the central tendency of the distribution is unchanged. Thus the continuous line in Fig. 18 gives the distribution of the averages of 100 individuals calculated from only 4 instead of from 20 records from each, the 4 being chosen at random from the 20. The broken line gives the distribution when all the 20 are used. Table XIX. gives the facts in figures. Table XX. and Fig. 19 give the distribution of the cost per pupil of supplies in 40 grammar schools for boys in New York City calculated from 2 and from 4 years' figures respectively. It is evident that one would not be much misled with respect to the general tendency of the group by taking the measure of the group from 4 records instead of that from 20, or even that from 2 instead of that from 4.

When the measure taken for an individual is his total ability, the measure of the group is, of course, a total distribution made up of all the separate individuals' distributions, each individual being given his proper share in determining this total distribution. In practice we rarely make up the total distribution of a trait in a group from adequate individual distributions, but use for each individual only a few measures. The result is very closely the same if the number of

TABLE XIX.

AVERAGE ERROR IN DRAWING A LINE TO EQUAL A 100-MM. LINE.

*A* = averages calculated from 20 trials for each individual.
*B* = averages calculated from 4 trials.

| Quantity: in tenths of millimeters. | Frequencies. *A.* | *B.* | Quantity: in tenths of millimeters. | Frequencies. *A.* | *B.* |
|---|---|---|---|---|---|
| −100 to −120 | 1 | 1 | +40 to +60 | 5 | 6 |
| −80 to −100 | 3 | 4 | +60 to +80 | 4 | 1 |
| −60 to −80 | 7 | 5 | +80 to +100 | 3 | 3 |
| −40 to −60 | 12 | 5 | +100 to +120 | | 0 |
| −20 to −40 | 17 | 18 | +120 to +140 | | 1 |
| 0 to −20 | 18 | 24 | Averages | −.72 mm. | −.46 mm. |
| 0 to +20 | 13 | 17 | Medians | −.889 mm. | −.584 mm. |
| +20 to +40 | 17 | 15 | | | |

TABLE XX.

COST PER PUPIL OF GENERAL SUPPLIES IN 40 BOYS' GRAMMAR SCHOOLS.

| Quantity, Dollars. | Frequency, 4 records used. | Frequency, 2 records used. | Quantity, Dollars. | Frequency, 4 records used. | Frequency, 2 records used. |
|---|---|---|---|---|---|
| 1.25–1.50 | 1 | 2 | 2.75 | 7 | 6 |
| 1.50–1.75 | 1 | 0 | 3.00 | 3 | 5 |
| 1.75, etc. | 2 | 5 | 3.25 | 1 | 1 |
| 2.00 | 6 | 7 | 3.50 | | 1 |
| 2.25 | 13 | 5 | Averages | 2.48 | 2.49 |
| 2.50 | 6 | 8 | Medians | 2.44 | 2.53 |

individuals is large. Thus the broken and continuous lines of Fig. 20 show practically the same fact, though the former gives the measure of the total ability of the group made up by putting together all the separate distributions from 20 trials each, while the latter gives the total ability made up by putting together only 4 from each. Table XXI. gives the facts in figures.

TABLE XXI.

ERRORS MADE BY 92 INDIVIDUALS IN DRAWING A LINE TO EQUAL A 100-MM. LINE. *A* GIVES THE DISTRIBUTION DUE TO 20 TRIALS FROM EACH INDIVIDUAL; *B* THAT DUE TO 4.* *B* IS RAISED TO AN EQUIVALENCE TO MAKE COMPARISON EASIER.

| Quantity. Error from standard, in mms. | Frequencies. *A.* | *B.* | Quantity. Error from standard, in mms. | Frequencies. *A.* | *B.* |
|---|---|---|---|---|---|
| 16 or less | 9 | 10 | + 3 | 187 | 177 |
| 14 | 13 | 0 | + 5 | 103 | 118 |
| 12 | 21 | 25 | + 7 | 89 | 108 |
| 10 | 62 | 64 | 9 | 47 | 44 |
| 8 | 84 | 94 | 11 | 33 | 39 |
| — 6 | 107 | 123 | 13 | 13 | 10 |
| — 4 | 217 | 207 | 15 | 6 | 10 |
| — 2 | 262 | 197 | 17 | 4 | 0 |
| 0 | 292 | 306 | 19 and over | 7 | 10 |
| + 1 | 256 | 237 | | | |

The determinations of the central tendency and variability of a measure of a group are made in just the same way as in the case of a measure of an individual, and the different measures of them have here the same characteristics. The formal and mathematical problem is identical whether we have varying records of one individual or varying individuals of one group, or varying records of many individuals in one group.

Starting, then, with the best measures of the individuals (for our purpose) that can be obtained, we put them together in a total distribution (allowing equal weight to each) and have the measure of the group. As in the case of individual measures, it is a safe rule never to replace this totality by any partial expressions of it until it is necessary. As in the case of an individual measured, the distributions may conceivably take all sorts of forms and be quite unrepresentable by any simple arithmetical constants.

But in point of fact the measurements of groups with which

* There were less than 20 trials in a few cases, hence the total numbers are not exactly 1840.

students of mental science have to deal do, in the case of most anatomical traits, of very many physiological traits, of many mental traits and of at least some institutional and social traits, show an approximation toward a distribution the variability of which is of such a nature as to justify one in regarding the members of the group as representatives clustering about a type, departures from which show a certain regularity. In other words, the statistical average or mode very often represents a real central tendency or type, and, the departures from it occurring in an orderly way, one or two figures can often represent the real clustering of individuals about a type.

In particular there is found very often a form of distribution (1) approximating the symmetrical, with its mode approximately at the average, so that both are nearly coincident with the median, and (2) characterized by a slow decrease in frequency for a certain distance above and below the mode, a more rapid decrease from then on for a way, and finally a slower decrease until the limits are reached. This description the reader will recognize as the description of a distribution approximating to the so-called normal distribution, that of a quantity determined by the action of a large number of independent causes equal in amount; in other words, that of the probability curve.

In so far as this uniformity in distributions does exist, we are freed from the necessity of devising a separate means of quantitative expression for each group measurement studied, and permitted to express it at least approximately in two figures, one telling the general tendency or type, the other the variability. The average, median and mode as measures of general tendency, and the average deviation, standard deviation, P. E., etc., as measures of variability, possess perhaps a wider and surer utility in the case of measures of groups than in the case of measures of individuals. The properties of the probability curve become of practical importance.

I have represented graphically in the following pages distributions of as many anatomical, physiological, mental, social and institutional traits as I could conveniently collect, drawing them so that a rough comparison with the surface of frequency of the probability integral could be made in each case.* The examination of these will

* The author will be much indebted to any of his readers who sends him the table of frequencies of any trait measured in any group, especially if the group is a large one. Such data must be at hand in any large hospital, school, psychological laboratory or gymnasium.

give a concrete and reasonably accurate notion of the frequency with which the measurement of a group is again and again approximately the same statistical problem.

In these figures (21 to 47) the continuous lines enclose the surface of frequency of the trait in question. The dotted lines give the surface which would be found if the distribution of the trait followed the type of the normal distribution, the probability surface. Where the actual distribution obviously does not follow this type even approximately, the dotted lines are omitted. The exact nature of the trait, the number of individuals and the source of the data in each case are given in the list that follows. When no source is stated the author is responsible for the original data.

FIG. 21. — Height of American adult men. In inches. *N* (number of cases) = 25,878. Drawn from the table given by Karl Pearson on page 385 of Vol. 186A of the *Philosophical Transactions of the Royal Society of London.* He quotes from J. H. Baxter, Medical Statistics of the Provost Marshal's General Bureau.

FIG. 22. — Weight of English adult men. In pounds. *N* = 5,552. Drawn from the table given in C. Roberts' 'Manual of Anthropometry'; appendix.

FIG. 23. — Cephalic Index (ratio of width to length of head) of modern Alt-Bayerische skulls. *N* = 900. Drawn from the table given by Karl Pearson in 'The Chances of Death.'

FIG. 24. — Length of male infants at birth. In inches. *N* = 451. Source the same as for Fig. 22.

FIG. 25. — Girth of chest, empty, of English army recruits. In inches. *N* = 675. Source the same as for Fig. 22.

FIG. 26. — Strength of arm pull. English adult men. Pull exerted as in drawing a bow. In pounds. *N* = 1497. Source the same as for Fig. 22.

Fig. 27. — Body temperature at the mouth in American women. *N* = 158. I am indebted for the original measures to Professor T. D. Wood, of Teachers College.

FIG. 28. — Heart rate (after vigorous exercise) in American students, young men 16 to 20. Number of beats per 60 seconds. *N* = 312. I am indebted for the original measures to Dr. G. L. Meylan, of Columbia University.

FIG. 29. — Reaction time of American college freshmen. Thousandths of a second. *N* = 252. I am indebted for the original measures to Dr. Clark Wissler, of the American Museum of Natural History.

FIG. 30 — Memory span for digits in American women students. Number of digits correctly written and correctly placed. *N* = 123.

FIG. 31. — Efficiency in perception of 12.5-year-old boys. Number of *A*'s marked in 60 seconds on a sheet of 13 lines of capital letters (see sample below). *N* = 312.

OYKFIUDBHTAGDAACDIXAMRPAGQZTAACVAOWLYXWABBTHJJANE
EFAAMEAACBSVSKALLPHANRNPKAZFYRQAQEAXJUDFOIMWZSAUC
GVAOABMAYDYAAZJDALJACINEVBGAOFHARPVEJCTQZAPJLEIQWN
AHRBUIAS

FIG. 32. — Efficiency in controlled association of 12.5-year-olds. Number of correct minus number of incorrect opposites of the following words written in 60 seconds: Good, outside, quick, tall, big, loud, white, light, happy, false, like, rich, sick, glad, thin, empty, war, many, above, friend. $N = 239$.

FIG. 33. — Accuracy of estimation of length in girls 13 to 15 years old.* Average variable error, in millimeters, in 30 attempts to draw a line equal to a 100-mm. line seen. $N = 153$.

FIG. 34. — Efficiency in complex perception of 12.5-year-old boys. Number of words containg *a* and *t* marked in 120 seconds in a sheet of words (see sample below). $N = 312$.

Dire tengo antipatia senores; esto seria necedad, porque hombre vale siempre tanto como otro hombre. Todas clases hombres merito; resumidas cuentas, sulpa suya vizxonde; pero dire sobrina puede contar dote viente cinco duros menos, tengo apartado; pardiez tamado trabajo atesorar-los para enriquecer estrano.

FIG. 35. — Ratio of attendance to enrollment in public schools of cities and towns of over 8,000 inhabitants in Ohio, Indiana, Illinois and Iowa. $N = 115$.

FIG. 36. — Wages of cotton operatives (in shillings per week). $N$ is large, but not given. The data are taken from Bowley's 'Elements of Statistics,' p. 96.

FIG. 37. — Age of graduation from American colleges. Men only taken. $N = 1,213$.

FIG. 38. — Cost per pupil of public school education in American cities of over 8,000 inhabitants. The cost is here taken per pupil actually present throughout the year. That is, the cost per pupil equals amount spent divided by average attendance. In dollars. $N = 465$. The amounts and average attendances are those given in the Report of the U. S. Commissioner of Education for 1901.

FIG. 39. — Wages of American workingmen per day. In cents. $N = 5,123$. The data are taken from Bowley's 'Elements of Statistics,' p. 120. He quotes them from a U. S. Senate report.

FIG. 40. — Figure 39 with a coarser grouping.

FIG. 41. — Ratio of attendance to enrollment in public schools of American cities of over 8,000 inhabitants. $N = 545$.

FIG. 42. — Incomes of American colleges for men and for both sexes. The five per cent. who in the year taken had incomes of over $150,000 are omitted. In thousands of dollars. $N = 438$.

FIG. 43. — Age at marriage of gifted American men. $N = 744$.

FIG. 44. — Frequency of divorces in different years after marriage. The cases after twenty years are undistributed by the compiler and are here given a probable distribution. $N = 109,960$. The data were taken from Karl Pearson's table, *Phil. Trans. of the Royal Society*, Vol. 186A, p. 395. He in turn quotes them from W. F. Wilcox, 'The Divorce Problem.'

FIG. 45. — Size of New England families, 1725–1800. The number of children born to women during twenty years or over of married life. $N = 163$.

FIG. 46. — Infant mortality in cities and towns of England and Wales. Number of deaths per 1,000 births. $N = 112$. Arranged from data given by Miss Clara Collet in the *Journal of the Royal Statistical Society*, June, 1898.

FIG. 47. — Frequency of death at different ages. After Karl Pearson, 'Chances of Death,' Vol. I., p. 27. $N$ is very large.

In figures 21 to 47, the limits to which the surface of frequency extends are shown by short vertical lines in those cases where the length of the columns of which it is composed is so small as to be unnoticeable. See, for instance, $l_1$ and $l_2$ in Fig. 21.

* The 13-, 14- and 15-year old girls did not differ as groups.

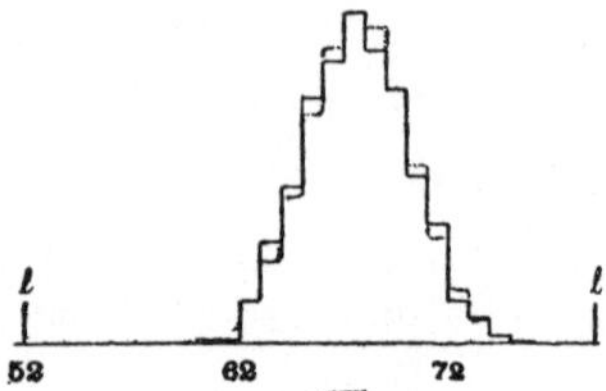


21

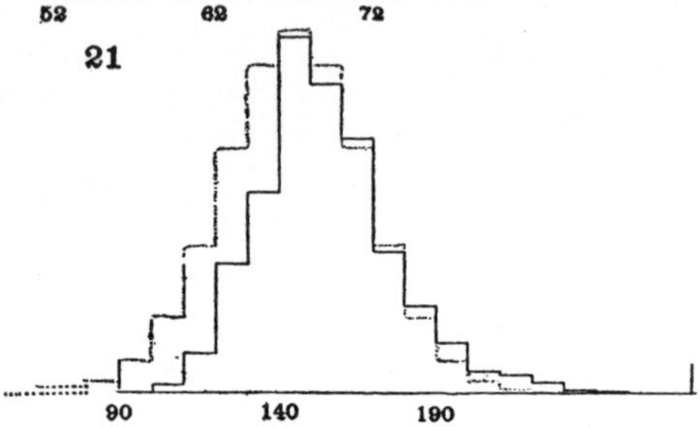


22

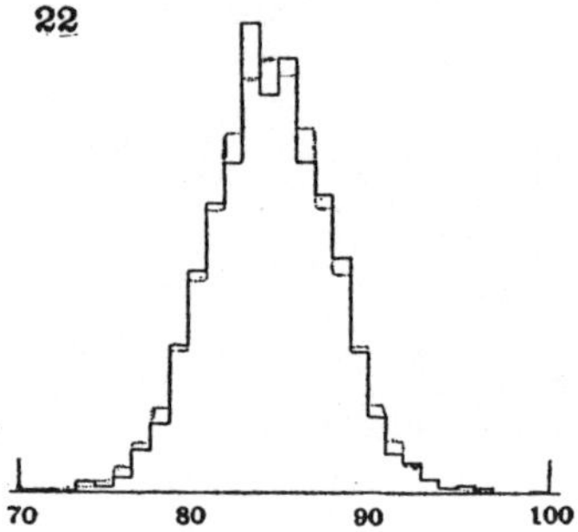


23

24

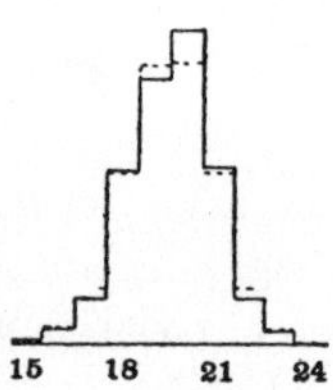


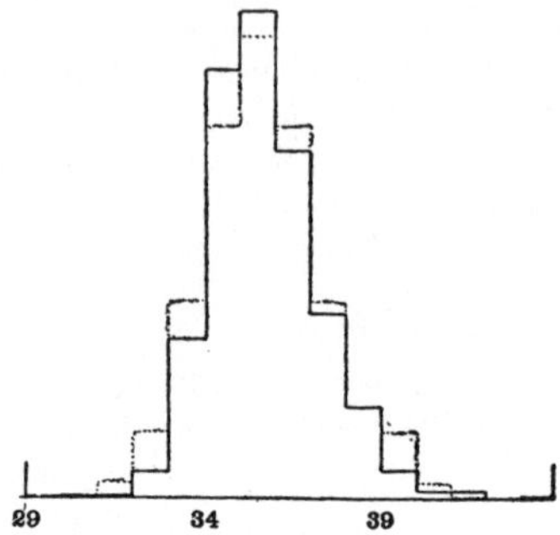


25

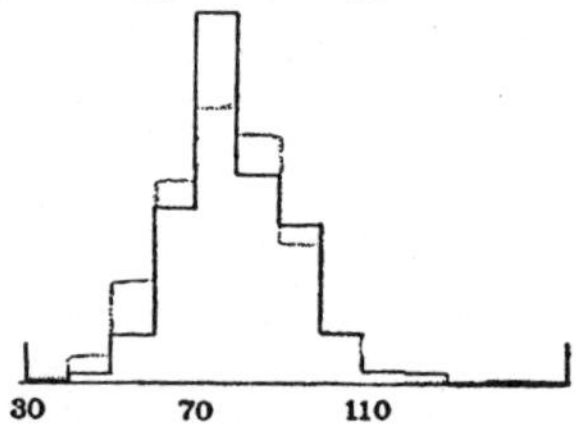


26

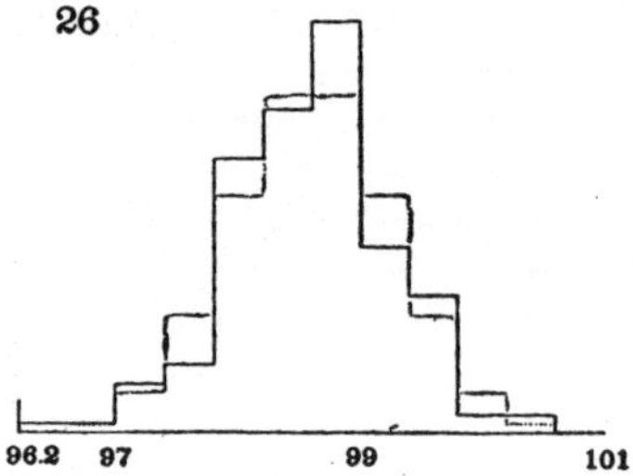


27

28

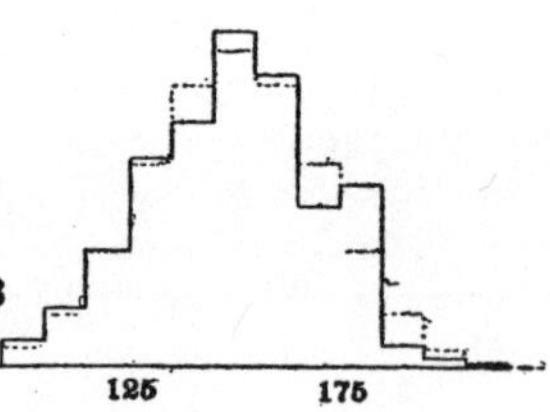


FIG. 21.—Height of men.
FIG. 22.—Weight of men.
FIG. 23.—Cephalic index.
FIG. 24.—Length of infants.
FIG. 25.—Girth of chest.
FIG. 26.—Strength of arm pull.
FIG. 27.—Body temperature.
FIG. 28.—Heart rate after exercise.

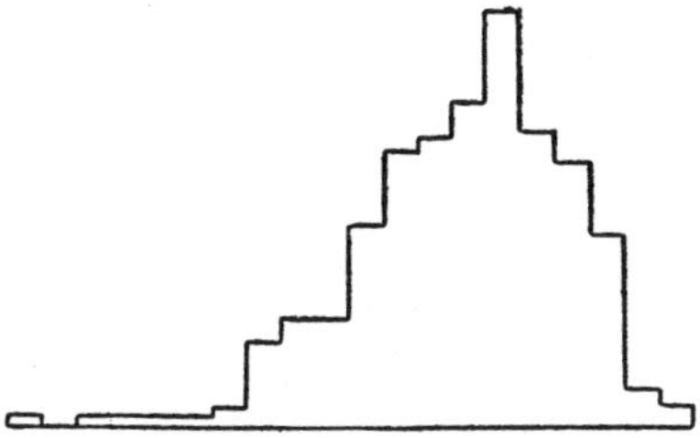

29

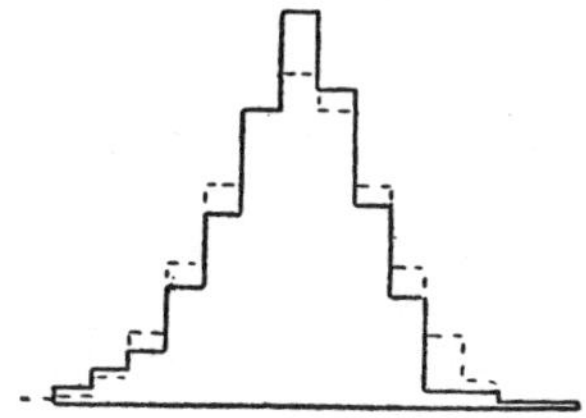

32

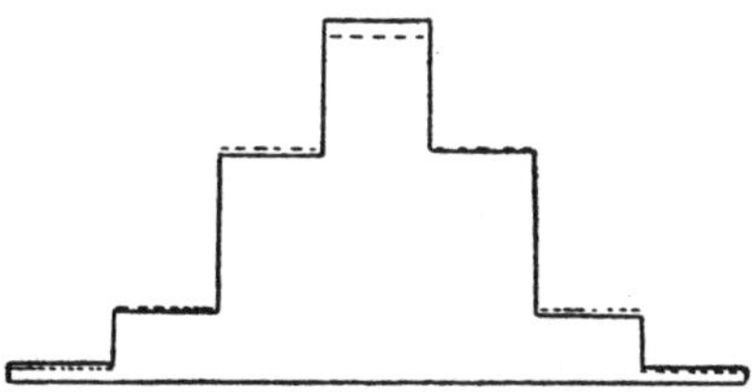

30

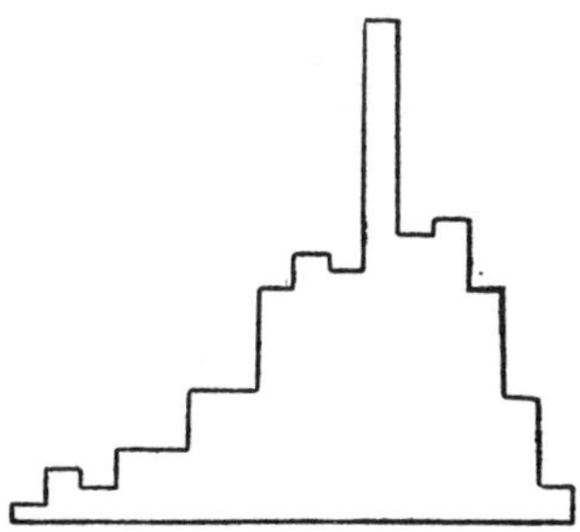

33

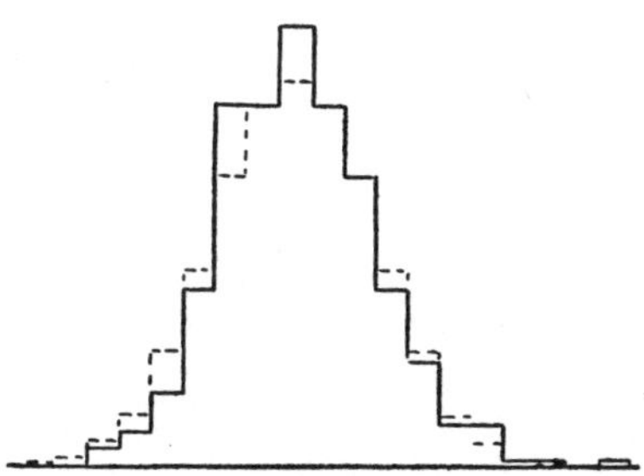

31

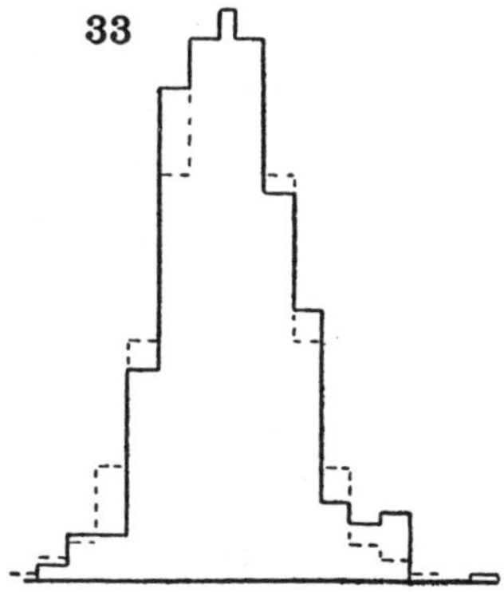

34

FIG. 29.—Reaction time.

FIG. 30.—Memory span for digits.

FIG. 31.—Efficiency in perception of As.

FIG. 32.—Efficiency in association of ideas.

FIG. 33.—Accuracy of estimation of length.

FIG. 34.—Efficiency in perception of words.

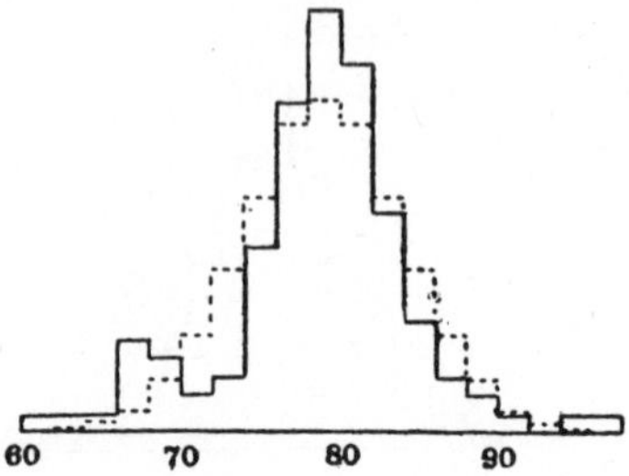


35

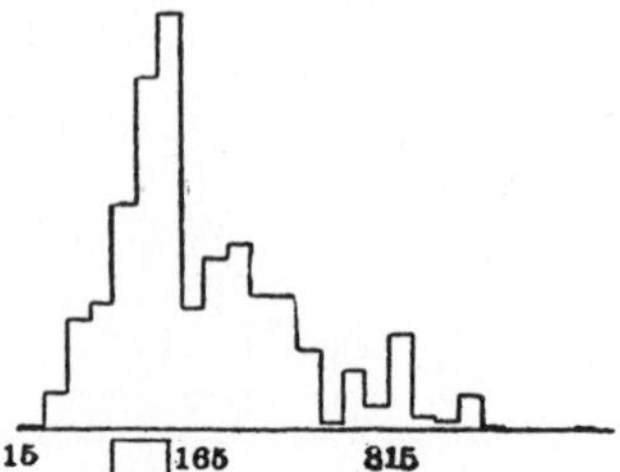


39

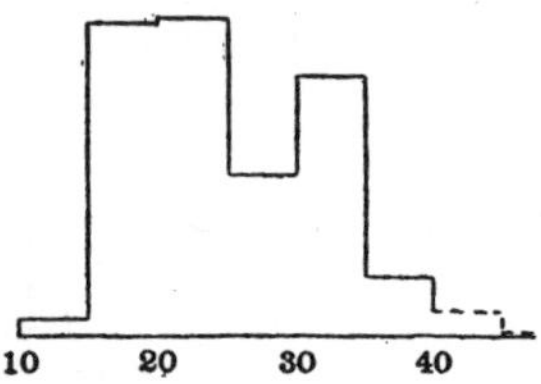


36

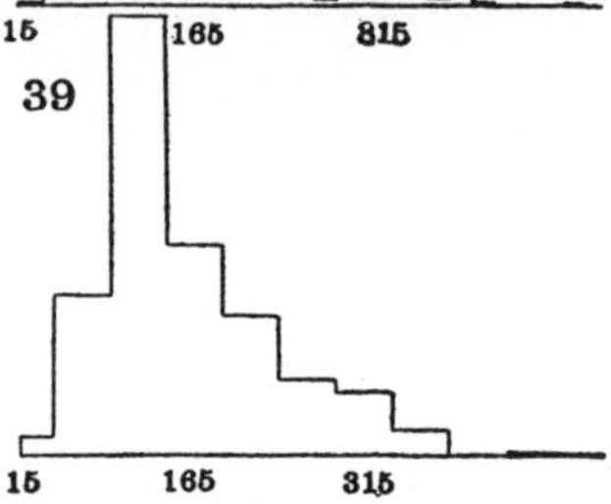


40

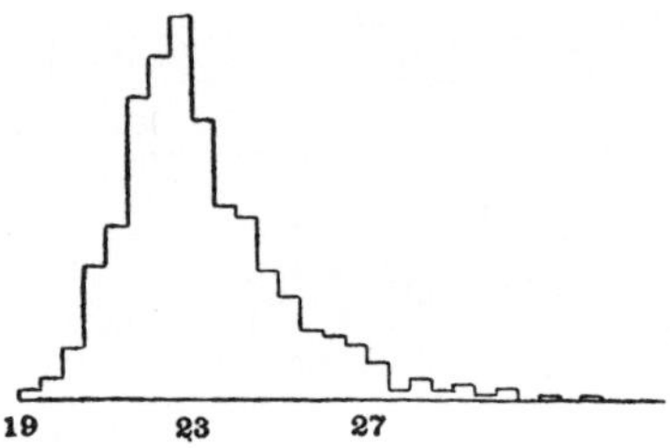


37

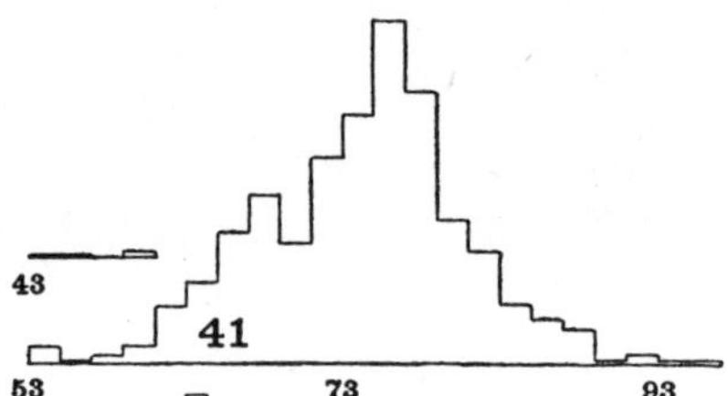


41

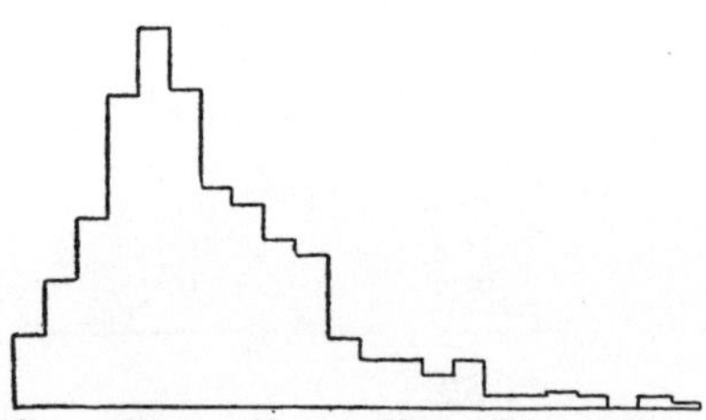


38

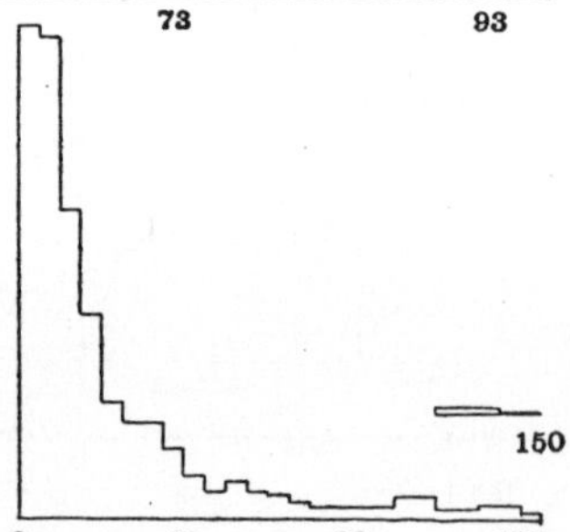


42

FIG. 35.—Ratio of school attendance to enrollment.
FIG. 36.—Wages of cotton operatives.
FIG. 37.—Age of graduation from college.
FIG. 38.—Cost per pupil of education.
FIG. 39.—Wages of American workingmen.
FIG. 40.—Wages of American workingmeu.
FIG. 41.—Ratio of school attendance to enrollment.
FIG. 42.—Incomes of colleges.

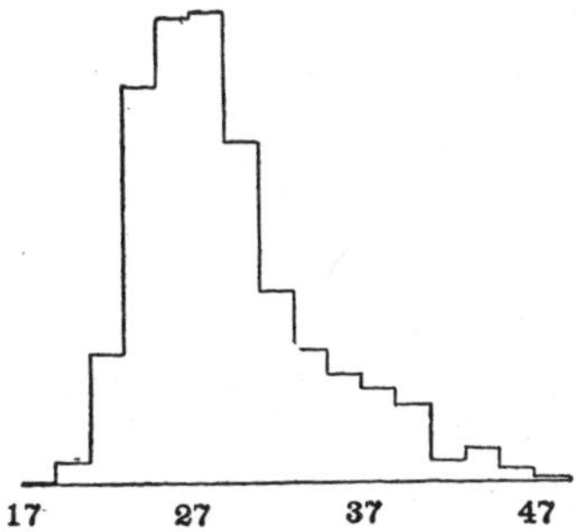


43

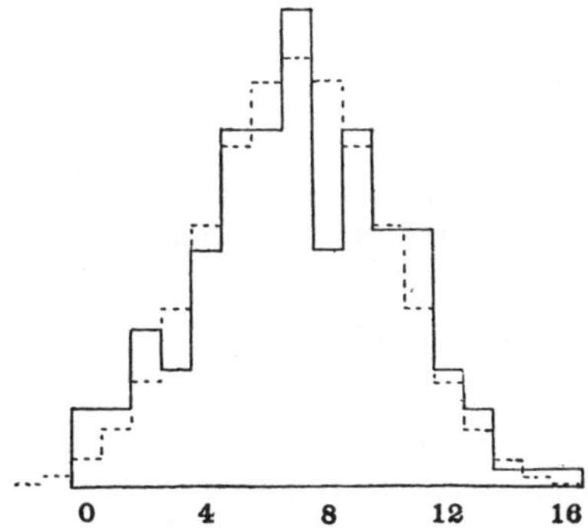


45

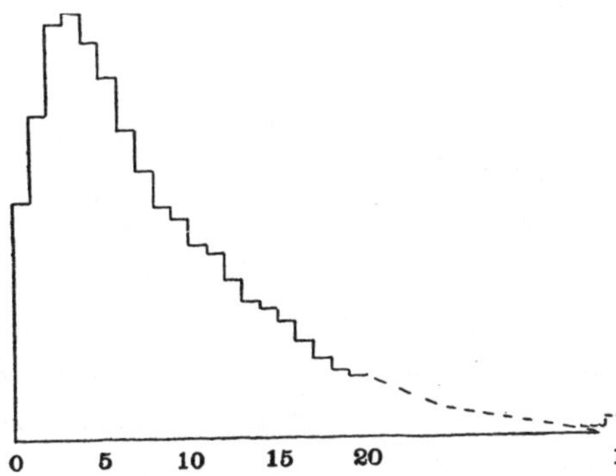


44

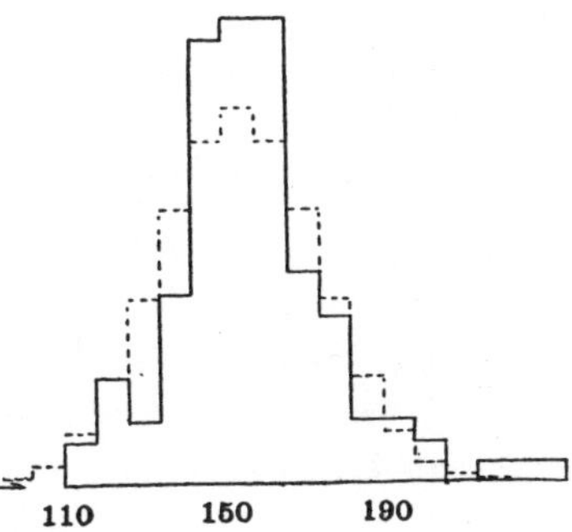


46

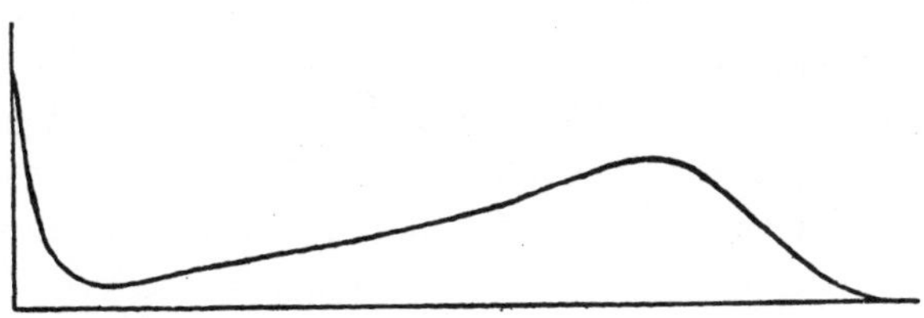

47

FIG. 43.—Age of marriage of gifted men.

FIG. 44.—Frequency of divorces at different dates after marriage.

FIG. 45.—Size of New Englaud families.

FIG. 46.—Infant mortality.

FIG. 47.—Frequency of death at different ages.

The quantitative expression of any group measurement in a few figures capable of treatment by ordinary statistical methods will depend upon (1) the considerations already explained in the case of individual measurements, and also upon our general information (2) about the group measured and (3) about the causes the action of which determines the quantity measured. A complete discussion of (2) and (3) is impossible because of the lack of data, and even such a survey as the inadequacy of the data permits would be far too intricate and obscure for the modest purposes of this book. All that will be attempted will be a rough statement of the facts about a group which are of most assistance in interpreting its surface of frequency, and a very elementary introduction to the study of the relation between the nature of the causes affecting a quantity and the quantity's distribution. The former will be the subject of the rest of this chapter; the latter will be given in Chapter V.

### *The Interpretation of the Form of a Surface of Frequency.*

It might appear reasonable to take the distribution obtained for any group at its face value. For instance, if in a measure of the scholarship of men one obtained a distribution like that represented in Fig. 48, it might appear reasonable to say that intellect was distributed in a very irregular manner and in such a way that there were no grades very far below the commonest condition, but that grades above it existed over such a range that the highest ranking person was ten times as far above the mode as the lowest ranking person was below it, and that the grades up near the highest were more common than those a little nearer the mode. Further consideration, however, might show that the infrequency of low grades was due to the fact that in our measurements we had tested only the better classes — had selected against the idiots, illiterates and incompetents; and that the apparently greater frequency of very high than of moderately high grades was due to our having measured some thousands of individuals from the better classes together with a hundred or so college graduates. Scholarship in general might really be distributed normally as shown in Fig. 49, and our result be due to the influence of selection and of mixing two species, untrained and trained men. On the other hand, if one obtained for scholarship a normal distribution, one could not be sure that in the natural group, men, scholarship was normally distributed unless these same factors of elimination and mixture were excluded. For example, if one got a normal distribution from measuring 13-year-old boys in the next to the last grammar-school grade, he could be practically sure that for all 13-year-old boys the distribution would *not* be normal. For the duller 13-year-old boys would not have reached that grade and the very bright ones would often have passed it. The actual distribution may be in part the result of the mixture of species or of selection.

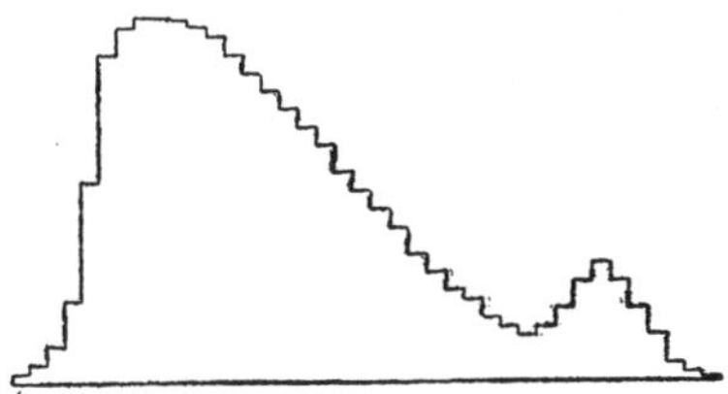

FIG.48.

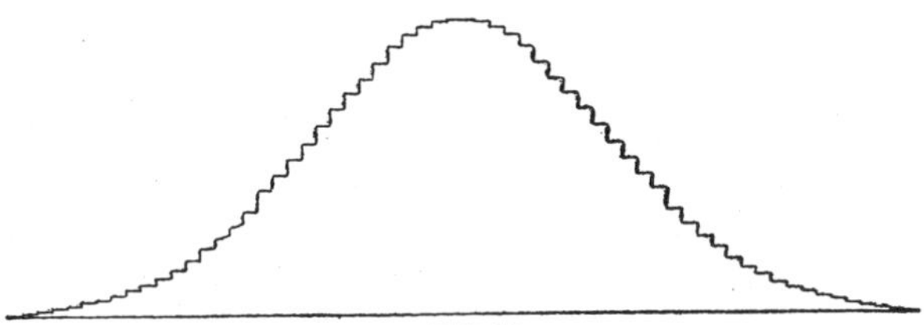

FIG.49.

FIG. 48. — An irregular distribution possibly due to artificial elimination and mixture of species in the course of the measurements.

FIG. 49.—A regular distribution.

## *Homogeneous and Mixed Groups.*

Homogeneity is in general not an absolute, but a relative, quality. A group of animals is homogeneous compared with a group of animals and plants mixed. A group of human beings is homogeneous compared with a group of men, dogs, worms and fishes. A group of college graduates is homogeneous compared with a group of college graduates, illiterates and idiots. Utter homogeneity would equal identity. We commonly mean by the homogeneity of any group with respect to any trait, such likeness amongst its members, with respect to the forces producing the trait, that there is no reason for separating them into several groups rather than leaving them in

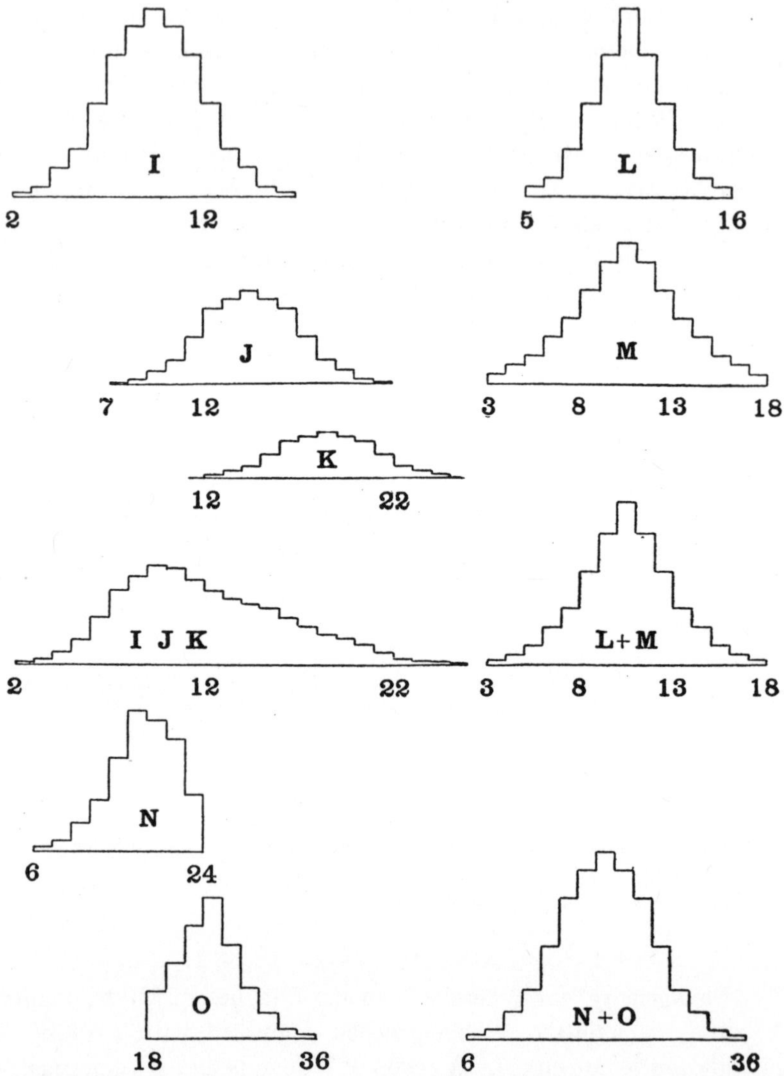


FIG. 50. — Showing six cases of the influence of combination upon the form of distribution, *viz:*

Two normal distributions, *A* and *B*, when combined, give a markedly bimodal distribution.

Two normal distributions, *C* and *D*, when combined, give a flattened distribution.

Four normal distributions, *E*, *F*, *G* and *H*, when combined, give a flattened and positively skewed distribution.

Three normal distributions, *I*, *J* and *K*, when combined, give a markedly skewed distribution.

Two distributions, *L* and *M*, of identical mode but differing variability, give, when combined, a form midway between the two.

Two distributions, *N* and *O*, one positively and the other negatively skewed, give, when combined, a normal distribution.

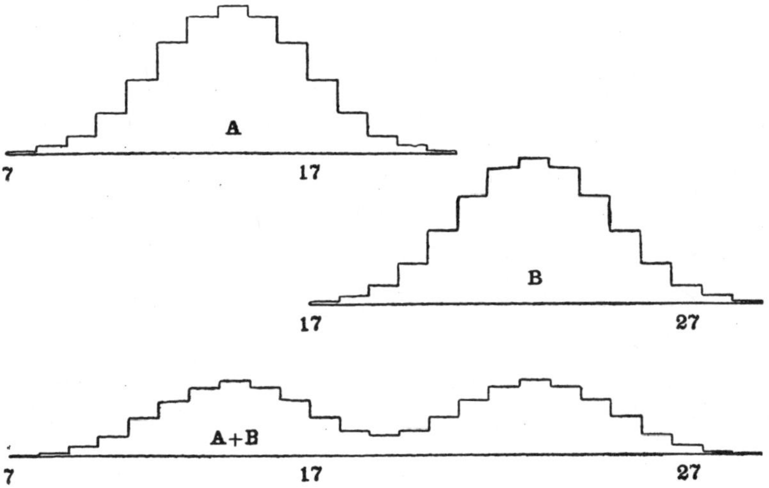
A
7
17
B
17
27
A+B
7
17
27

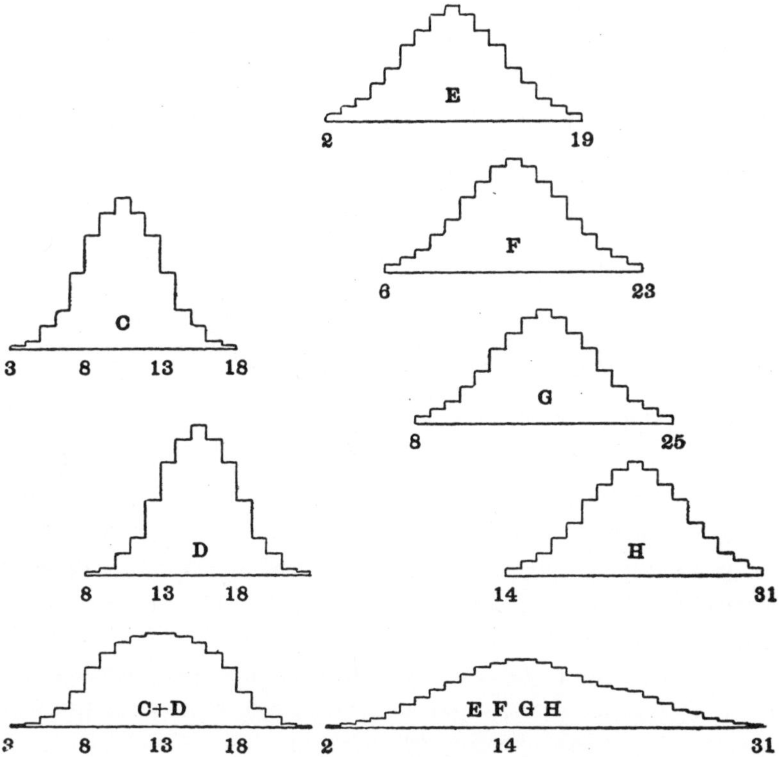
E
2
19
F
6
23
C
3
8
13
18
G
8
25
D
8
13
18
H
14
31
C+D
3
8
13
18
E F G H
2
14
31

one. Thus the group 'a species' of the zoölogist or botanist is homogeneous with respect to its anatomy. Thus the group 'children of the same race, sex and age' is probably homogeneous with respect to the trait 'maturity.' Thus the group 'wages of unskilled laborers

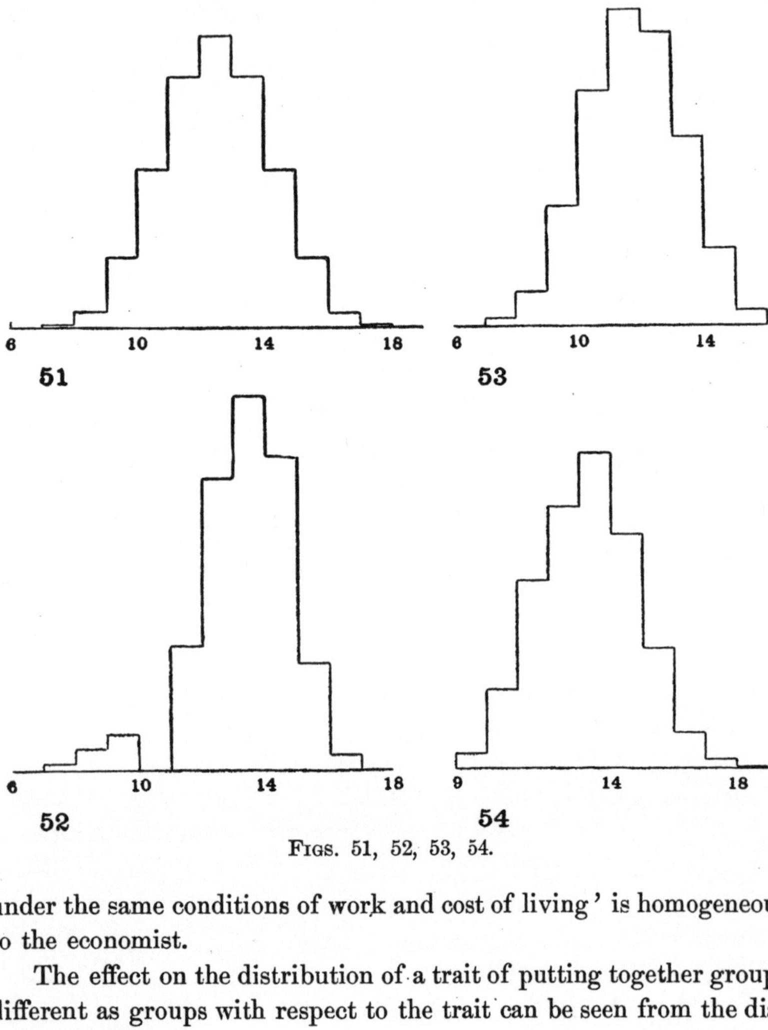


FIGS. 51, 52, 53, 54.

under the same conditions of work and cost of living' is homogeneous to the economist.

The effect on the distribution of a trait of putting together groups different as groups with respect to the trait can be seen from the diagrams of Fig. 50.

It is obvious, in general, that given any form of distribution, it might be accounted for, so far as the bare fact of its existence went, by any one of a practically infinite number of different compoundings of groups. The mere form of distribution does not itself tell. Recourse must be had to a study of the real facts about the group.

I shall consider further only the case of the compounding of two or more groups, each of which by itself shows approximately normal distribution, which differ in respect to the amount of the trait. It is clear from the diagrams that the result on the form of distribution of the total group will be multimodality or a flattening of the top of the surface of frequency at some point. If one has reason to believe that the trait he is studying would in a homogeneous group show normal distribution, the existence of such multimodality or flattening may properly lead him to suspect the mixture of two groups or species and to examine the cases with a view to separating them into more homogeneous groups.

One special case of importance is that where the total group is a compound of a very large number of groups so differing that their central tendencies form approximately an arithmetical series. Such total groups would be, for instance, measurements of children eight to twelve years of age in some physical or mental trait subject to growth, or of teachers' salaries over a period of years during which there was a steady rise in values. The death-rate for children under a year reckoned on the last thirty years' records in 100 cities would be a mixture of thirty different groups.

### *Selected Groups.*

Only very infrequently does the measurement of any trait in a group include all the members of a group. It is, on the contrary, the result of measurements of relatively few sample individuals. These represent the group as a whole justly only in so far as they include the same percentage of each grade of ability in the group. Suppose the real distribution to be as given in Fig. 51. If 20 per cent. of each grade are taken, the form of distribution, of course, remains as before. If 20 per cent. of grade 1, 18 per cent. of grade 2, 16 per cent. of grade 3, and so on, are taken, the form becomes that of Fig. 52. If the per cents. taken are in order 20, 15, 10, 5, 0, 5, 10, 15, 20, 15, 10, 5, 0, the form of distribution becomes that of Fig. 53. If the per cents. taken are in order 0, 0, 0, 10, 20, 30, 40, 50, 60, 70, 80, 90 and 100, the form of distribution becomes that of Fig. 54.

In general, it can easily be shown that by the right combinations of selections from a group, a group with any form of distribution can be derived, no matter what the form of distribution of the trait in the original group was.

Selection may occur (1) as a result of natural forces upon a group, or (2) as the result of unproportional sampling by the measurer.

The group, living human beings 40 years old, is thus the selection by natural forces from the group, all human beings born 40 years ago, a selection, to some extent at least, of the physically more vigorous, morally less murderous, and so on.

The group, seventeen-year-old boys measured in school, is a selection from all boys seventeen years old, due to the measurer's willingness to take boys not absolutely at random, but as found conveniently. The selection is, to some extent at least, of the more ambitious and gifted intellectually.

The influence of nature in changing the distribution of a trait in a group by selecting for survival on the basis of the trait's amount is one of the most important topics for science, but does not need further mention here. The influence of circumstances in providing the student with a set of selected samples the distribution of which is unlike that of the total group the student takes them to represent, is, on the other hand, the most important cause of the majority of statistical fallacies in the mental sciences, and requires discussion here and in another connection later.

Although any form of frequency surface may be derived from any other by the proper method of selection of cases, and although, consequently, from the actual form of a surface of frequency nothing can be concluded concerning the group from which it represents a selection unless the method of selection is known, yet certain appearances may well serve to awaken suspicion and guide the student to interpret the measurements. In particular, skewness is so often connected with picking for study extreme cases of a group which as a whole would give approximately normal distribution, that it is certainly advisable always, when confronted by a group measure showing skew distribution, to ascertain whether the group is not a partial picking from a normally distributed total group.

The reader who has carefully attended to the numerous theoretical reservations and cautions of this chapter will now be able to use, and not abuse, the general practical advice to which it leads, which is:

If for any reason you have to make an hypothesis about the form of distribution of any trait in the absence of the facts, the most likely one in the case of anatomical, physiological and mental traits is that the form will be something like the normal probability surface. The probable error of any such hypothesis is least in the case of anatomical traits. Prediction of the form of distribution of economic traits is very insecure. The interpretation of any ascertained form of distribution is difficult, but may prove very instructive.

If in dealing with group measurements you can, *without violating any known fact*, use the hypothesis that in a homogeneous group not subject to selection on the basis of the trait in question, any mental trait due to natural as opposed to artificial causes, is distributed approximately normally, do so.

The normal surface of frequency (which is that of a quantity due to the chance combinations of $n$ causes, all equal and independent, when $n$ is infinitely large) is, as stated on page 36, the surface enclosed by the normal probability curve,

$$\left( Y = Pe^{\frac{-x^2}{2npq}} \quad \text{or} \quad y = e^{-x^2}, \right.$$

or some specialized form, as

$$\left. y = \frac{1}{\mu\sqrt{2\pi}} e^{\frac{-x^2}{2\mu^2}} \right)$$

and the abscissa or base line on which $x$ is scaled.

In this form of distribution the Average, Median and Mode coincide, for $y$ is the same for any given $-x$ as for the same $+x$, and is greatest when $x = 0$. Constant relations hold between the different measures of variability, *viz*:

$$\sigma = 1.25331 \text{ A. D.}$$
$$\sigma = 1.4825 \text{ P. E.}$$
$$\text{A. D.} = .7979\,\sigma$$
$$\text{A. D.} = 1.1843 \text{ P. E.}$$
$$\text{P. E.} = .6745\,\sigma$$
$$\text{P. E.} = .8453 \text{ A. D.}$$

Between Av. $-\sigma$ and Av. $+\sigma$ are 68.2 per cent. of the cases.

" Av. = A. D. and Av. + A. D. are 59.5 per cent. of the cases.

" Av. − P. E. " Av. + P. E. " 50 " " " " "

The frequencies of different deviations from the mode (or average or median) are given in gross in Tables XXII. and XXIII. Detailed tables will be given later.

### TABLE XXII.

FREQUENCIES OF DEVIATIONS ABOVE THE MODE IN A NORMAL SURFACE OF FREQUENCY IN TERMS OF A. D. THE FIGURES CAN BE USED IDENTICALLY FOR MINUS DEVIATIONS.

| | | | | | | | | |
|---|---|---|---|---|---|---|---|---|
| Between | + 0 | and | + .2 A. D. | are | 6.3 | per cent. | of the | cases. |
| " | + .2 | " | + .4 | " | 6.2 | " | " | " |
| " | + .4 | " | + .6 | " | 5.9 | " | " | " |
| " | + .6 | " | + .8 | " | 5.4 | " | " | " |
| " | + .8 | " | + 1.0 | " | 5.0 | " | " | " |
| " | + 1.0 | " | + 1.2 | " | 4.3 | " | " | " |
| " | + 1.2 | " | + 1.4 | " | 3.7 | " | " | " |
| " | + 1.4 | " | + 1.6 | " | 3.1 | " | " | " |
| " | + 1.6 | " | + 1.8 | " | 2.6 | " | " | " |
| " | + 1.8 | " | + 2.0 | " | 2.0 | " | " | " |
| " | + 2.0 | " | + 2.2 | " | 1.5 | " | " | " |
| " | + 2.2 | " | + 2.4 | " | 1.2 | " | " | " |
| " | + 2.4 | " | + 2.6 | " | 0.9 | " | " | " |
| " | + 2.6 | " | + 2.8 | " | 0.6 | " | " | " |
| " | + 2.8 | " | + 3.0 | " | 0.5 | " | " | " |
| " | + 3.0 | " | + 3.2 | " | 0.26 | " | " | " |
| " | + 3.2 | " | + 3.4 | " | 0.21 | " | " | " |
| " | + 3.4 | " | + 3.6 | " | 0.13 | " | " | " |
| " | + 3.6 | " | + 3.8 | " | 0.07 | " | " | " |
| " | + 3.8 | " | + 4.0 | " | 0.06 | " | " | " |
| " | + 4.0 | " | + 4.2 | " | 0.03 | " | " | " |
| " | + 4.2 | " | + 4.4 | " | 0.02 | " | " | " |
| " | + 4.4 | " | + 4.6 | " | 0.01 | " | " | " |
| " | + 4.6 | " | + ∞ | " | 0.01 | " | " | " |

### TABLE XXIII.

FREQUENCIES OF PLUS DEVIATIONS OR OF MINUS DEVIATIONS. IN TERMS OF $\sigma$.

| | | | | | | | | |
|---|---|---|---|---|---|---|---|---|
| Between | 0 | and | $.2\sigma$ | are | 7.93 | per cent. | of the | cases. |
| " | .2 | " | .4 | " | 7.61 | " | " | " |
| " | .4 | " | .6 | " | 7.04 | " | " | " |
| " | .6 | " | .8 | " | 6.24 | " | " | " |
| " | .8 | " | 1.0 | " | 5.32 | " | " | " |
| " | 1.0 | " | 1.2 | " | 4.36 | " | " | " |
| " | 1.2 | " | 1.4 | " | 3.43 | " | " | " |
| " | 1.4 | " | 1.6 | " | 2.59 | " | " | " |
| " | 1.6 | " | 1.8 | " | 1.89 | " | " | " |
| " | 1.8 | " | 2.0 | " | 1.32 | " | " | " |
| " | 2.0 | " | 2.2 | " | 0.88 | " | " | " |
| " | 2.2 | " | 2.4 | " | 0.57 | " | " | " |
| " | 2.4 | " | 2.6 | " | 0.35 | " | " | " |
| " | 2.6 | " | 2.8 | " | 0.22 | " | " | " |
| " | 2.8 | " | 3.0 | " | 0.12 | " | " | " |
| " | 3.0 | " | 3.2 | " | 0.06 | " | " | " |
| " | 3.2 | " | 3.4 | " | 0.04 | " | " | " |
| " | 3.4 | " | 3.6 | " | 0.02 | " | " | " |
| " | 3.6 | " | ∞ | " | 0.01 | " | " | " |

# CHAPTER V.

## THE CAUSES OF VARIABILITY AND THE APPLICATION OF THE THEORY OF PROBABILITY TO MENTAL MEASUREMENTS.

THE varying measures of one individual's performances and the varying measures of the individuals in a group were found in Chapters III. and IV. to be often distributed approximately after the fashion of the surface of frequency enclosed by the probability curve and its abscissa. In those chapters brief mention was made of the properties of this surface or type of distribution, acquaintance with which is a great assistance to convenient handling of mental measurements. The recognition of this type of frequency surface, the appreciation of its meaning and that of certain common departures from it, and the use of tables derived from the probability integral in calculations of measurements of traits approximately normally distributed, are all possible, at least to the moderate degree required for ordinary statistical work, without any knowledge of the abstract principles involved. But such knowledge is well worth obtaining for the sake of the additional insight into the meaning of concrete facts thereby given, and even merely for the sake of the additional facility in the use and construction of tables and the common formulæ. The present chapter will, therefore, contain a very simple introduction to the study of the applications of the mathematics of probability to the theory of the distribution of mental traits. From it the student may proceed to the study itself with the aid of the references given at the end of the chapter. The chapter will also introduce the student to the more general problem of the relation which the nature of the causes determining the amount of a trait hold to the trait's distribution.

Let us begin with the consideration of a quantity which is dependent on the action of one cause which is as likely to occur as not, and call the cause $a$. For example, $a$ may be the action of John's father in giving him a Christmas gift of a dollar.

The condition of affairs resulting will be, of course, no action or $a$. The quantity in question, John's Christmas money, will be 0 or $1.00. Its distribution will be

| Quantity. Dollars. | Frequency. Per cent. |
|---|---|
| 0 | 50 |
| 1 | 50 |

Its surface of frequency will be a rectangle, composed of two rectangles of equal base and altitude.

Suppose now that two causes contribute to determine the quantity, *a* and *b*, the possible actions of John's father and mother, and that all combinations of these causes are equally likely. The condition of affairs resulting will be, then, no action, *a*, *b* or *ab*, all being equally likely. If now $a =$ a gift of $1.00 and *b* likewise, the quantity in question, John's Christmas money, will be 0, $1.00, $1.00 or $2.00. Its distribution will be

| Quantity. Dollars. | Frequency. Per cent. |
|---|---|
| 0 | 25 |
| 1 | 50 |
| 2 | 25 |

Its surface of frequency is that shown in Fig. 55. If the conditions are kept the same but the number of causes increased to three, the condition of affairs will be, no action, *a*, *b*, *c*, *ab*, *ac*, *bc*, or *abc*. If as before $a = b = c$ in magnitude, John will get $2.00 as often as $1.00 and three times as often as nothing or $3.00.

The surface of frequency of the quantity, John's Christmas money, will be four rectangles, as shown in Fig. 56.

Keeping all the conditions the same, let the number of causes be increased to 4, then to 5, and then to 6. The condition of affairs in each case and the resulting distribution-schemes and surfaces of frequency are given in Tables XXIV., XXV. and XXVI., and Figs. 57, 58 and 59.

TABLE XXIV.

Combinations of 4 Causes, *a*, *b*, *c* and *d*.

| | | | | | | Value in Dollars. | Probable Frequency. |
|---|---|---|---|---|---|---|---|
| 0 | | | | | | 0 | 1 |
| *a*, | *b*, | *c*, | *d* | | | 1.00 | 4 |
| *ab*, | *ac*, | *ad*, | *bc*, | *bd*, | *cd* | 2.00 | 6 |
| *abc*, | *abd*, | *acd*, | *bcd* | | | 3.00 | 4 |
| *abcd* | | | | | | 4.00 | 1 |

## TABLE XXV.

COMBINATIONS OF 5 CAUSES, *a*, *b*, *c*, *d* AND *e*.

| | Value in Dollars. | Probable Frequency. |
|---|---|---|
| 0 | 0 | 1 |
| *a*, *b*, *c*, *d*, *e* | 1.00 | 5 |
| *ab*, *ac*, *ad*, *ae*, *bc*, *bd*, *be*, *cd*, *ce*, *de* | 2.00 | 10 |
| *abc*, *abd*, *abe*, *acd*, *ace*, *ade*, *bcd*, *bce*, *bde*, *cde* | 3.00 | 10 |
| *abcd* *abce*, *abde*, *acde*, *bcde* | 4.00 | 5 |
| *abcde* | 5.00 | 1 |

## TABLE XXVI.

COMBINATIONS OF 6 CAUSES, *a*, *b*, *c*, *d*, *e* AND *f*.

| | Value in Dollars. | Probable Frequency. |
|---|---|---|
| 0 | 0 | 1 |
| *a*, *b*, *c*, *d*, *e*, *f* | 1.00 | 6 |
| *ab*, *ac*, *ad*, *ae*, *af*, *bc*, *bd*, *be* | | |
| *bf*, *cd*, *ce*, *cf*, *de*, *df*, *ef* | 2.00 | 15 |
| *abc*, *abd*, *abe*, *abf*, *acd*, *ace*, *acf* | | |
| *ade*, *adf*, *aef*, *bcd*, *bce*, *bcf*, *bde* | | |
| *bdf*, *bef*, *cde*, *cdf*, *cef*, *def* | 3.00 | 20 |
| *abcd*, *abce*, *abcf*, *abde*, *abdf* | | |
| *abef*, *acde*, *acdf*, *acef*, *adef* | | |
| *bcde*, *bcdf*, *bcef*, *bdef*, *cdef* | 4.00 | 15 |
| *abcde*, *abcdf*, *abcef*, *abdef* | | |
| *acdef*, *bcdef* | 5.00 | 6 |
| *abcdef* | 6.00 | 1 |

## TABLE XXVII.

COMBINATIONS OF 10, 15 AND 20 CAUSES.

| Quantity. Dollars. | Frequency in case. Of 10. | Of 15. | Of 20. |
|---|---|---|---|
| 0 | 1 | 1 | 1 |
| 1 | 10 | 15 | 20 |
| 2 | 45 | 105 | 200 |
| 3 | 120 | 455 | 1,080 |
| 4 | 210 | 1,365 | 4,505 |
| 5 | 252 | 3,003 | 14,944 |
| 6 | 210 | 5,005 | 38,370 |
| 7 | 120 | 6,435 | 77,420 |
| 8 | 45 | 6,435 | 125,970 |
| 9 | 10 | 5,005 | 167,960 |
| 10 | 1 | 3,003 | 184,756 |
| 11 | | 1,365 | 167,960 |
| 12 | | 455 | 125,970 |
| 13 | | 105 | 77,420 |
| 14 | | 15 | 38,370 |
| 15 | | 1 | 14,944 |
| 16 | | | 4,505 |
| 17 | | | 1,080 |
| 18 | | | 200 |
| 19 | | | 20 |
| 20 | | | 1 |
| 21 | | | |

It is apparent that the surface of frequency of a quantity dependent upon the action of causes equal in magnitude, any combination of which is equally probable, tends, as the number of these causes becomes great, to approach the type we often find in the case of anatomical traits. This is emphasized by Table XXVII. and Figs. 60, 61 and 62, which give the results in our illustration if the number of causes is increased to 10, 15 and 20 respectively. When the number of causes is very, very great the result is the normal probability surface (Fig. 63).

The normal type of distribution may therefore be expected in the case of the different performances or measures of an individual in the same trait, if any one of his performances in the trait is due to the action of some one combination from a large number of causes of equal magnitude which are independent of each other, so that any combination is as likely to occur as any other; may be expected in the case of the different measures of individuals in a group, if the tendency of any individual in the trait is due to the action of some one combination, characteristic of his make-up, from such a large number of causes. If, that is, we think of any single act of a person as a result of a chance combination from amongst a number of causes which determine acts of that sort characteristic of him, we shall expect his different manifestations of the trait of which that act is a sample to be normally distributed; so also, if we think of the quantity of a trait in any single individual of a group as a result of a chance combination from amongst a number of causes characteristic of the group as a whole which determine that trait, we shall expect the manifestations of that trait by the group of which he is a sample to be normally distributed.

The clause 'so that any combination is as likely to occur as another' and its synonymous phrase 'a chance combination from amongst' need some explanation. They refer to the fact that the causes must be independent of each other if the distribution of the trait is to be normal. The need of this condition will be apparent from a concrete illustration.

Suppose that in our previous case of John's Christmas money the six causes *a*, *b*, *c*, *d*, *e* and *f* were as before, except that no action was barred out, and that if *a* acted *b* and *c* must also, and *d*, *e* and *f* could not; while if *d* acted *e* and *f* must, but *a*, *b* and *c* could not.

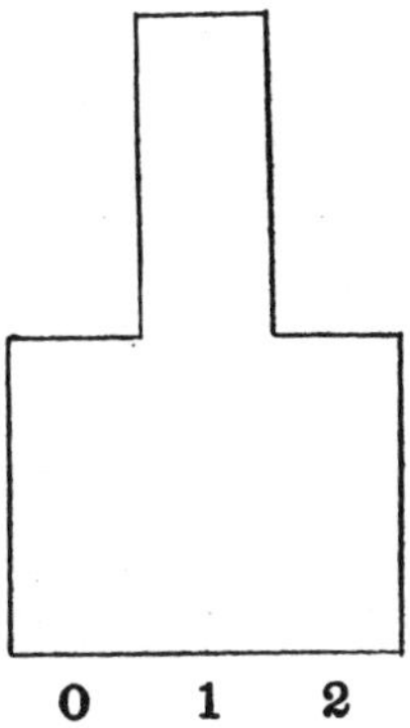


55

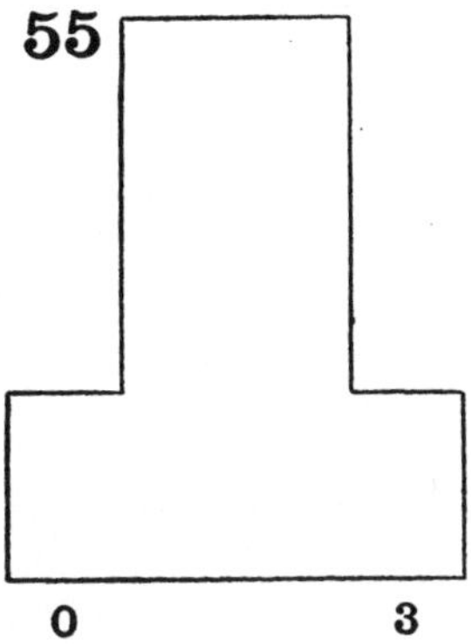


56

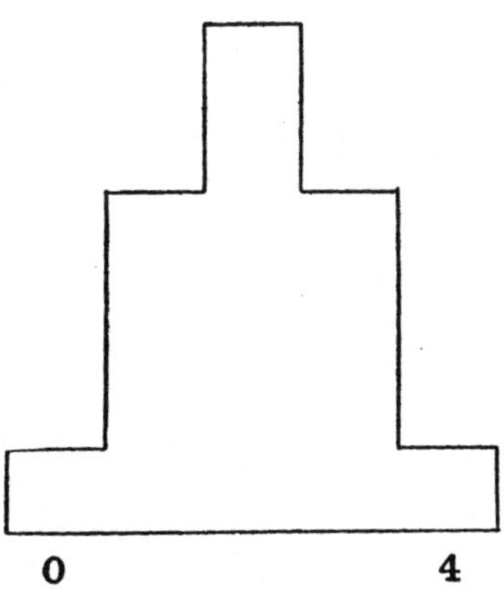


57

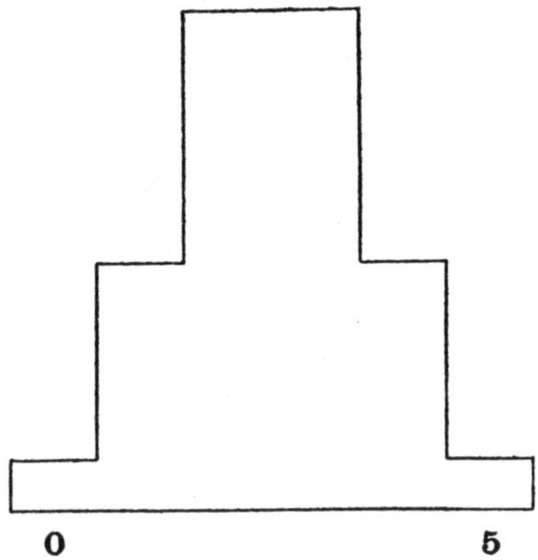


58

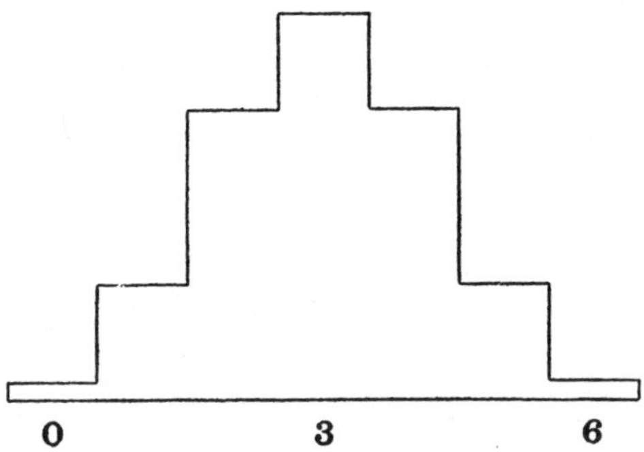


59

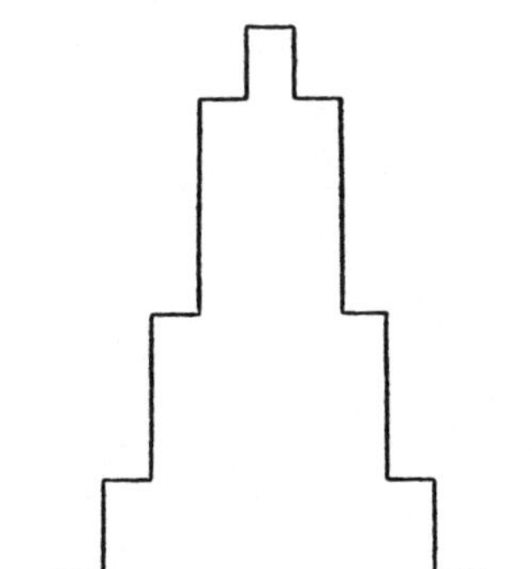

60

FIGS. 55, 56, 57, 58, 59, 60.

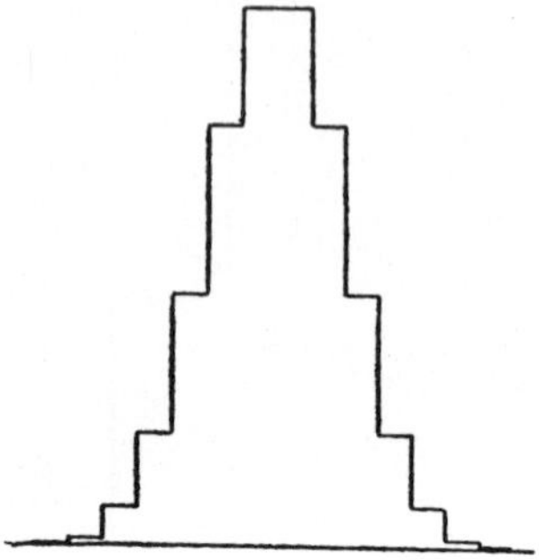

61

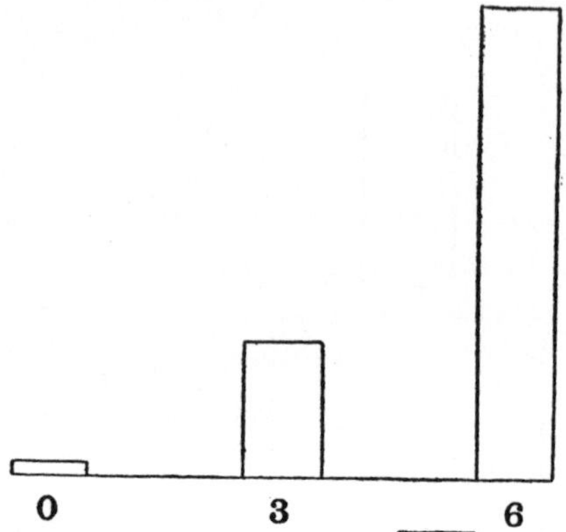


64

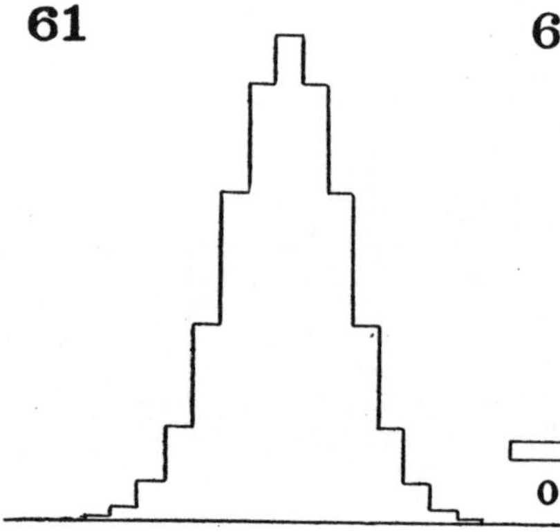

62

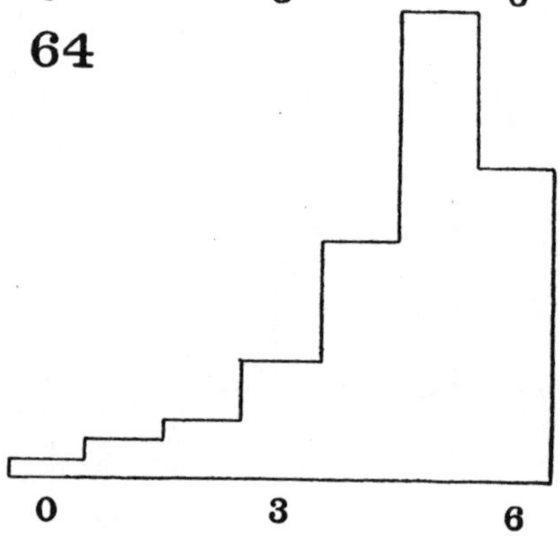


65

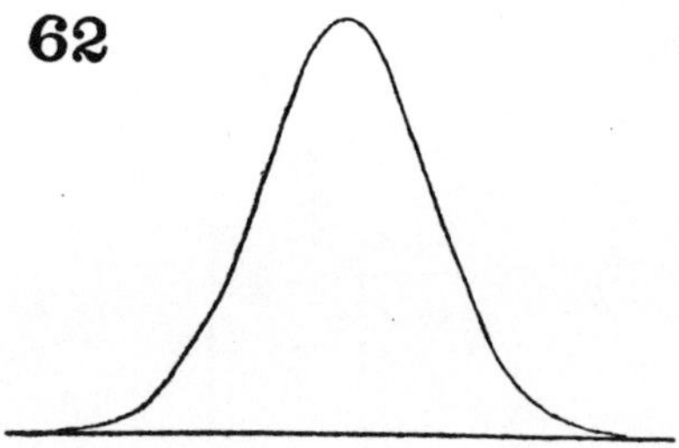

63

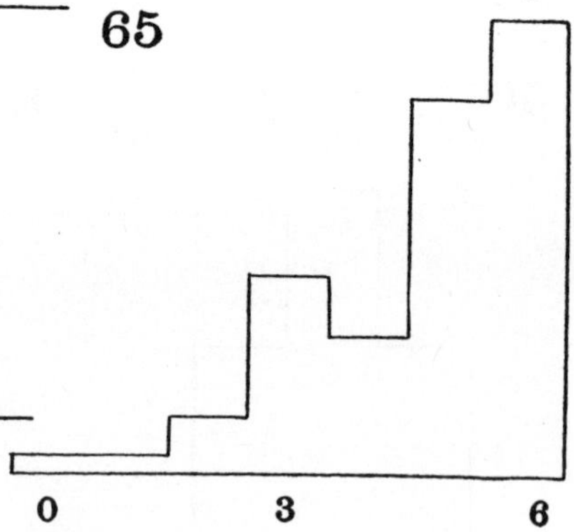


66

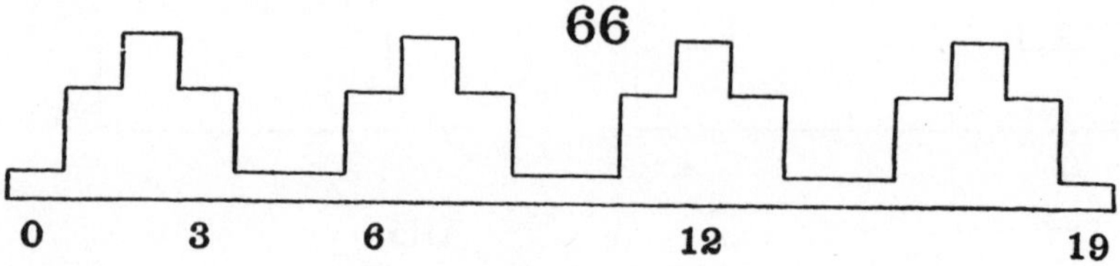


67

FIGS. 61, 62, 63, 64, 65, 66, 67.

Imagine, for instance, that it was agreed to take turns in preventing a penniless Christmas; that the father agreed to give his dollar if the mother and sister would always join with him and the grandfather, grandmother and brother would keep their money to themselves, while the grandfather agreed to give his dollar upon the condition that he be joined by grandmother and brother and that father, mother and sister refrain. The condition of affairs then could only be *abc* or *def* instead of the range of possibility of the illustration in its first form. Although there are six causes, the result is as if there were only one, and that always operative.

Suppose the presence of *a* or *b* or *c* to always cause that of the other two of the three, and similarly for the presence of *d*, *e* or *f*. This means that whenever cause *a* appears it adds to itself *b* and *c*, whenever *b* appears it adds to itself *a* and *c*, and so on. Every condition in Table XXVI. with *a* or *b* or *c* in it must then become *abc*; every condition with *d* or *e* or *f* must become *def*; every condition with one from the *abc* and one from the *def* group must become *abcdef*. Thus the condition of affairs would be, instead of that in Table XXVI., the following: no action, 1; *abc*, 7; *def*, 7; *abcdef*, 49.

The distribution would then be (as shown in Fig. 64):

| Quantity. Dollars. | Frequency. |
|---|---|
| 0 | 1 |
| 3 | 14 |
| 6 | 49 |

Suppose the presence of *a* to imply always that of *c*, *d*, *e* and *f*, the presence of *b* to imply always that of *d*, *e* and *f*, the presence of *c* to imply that of *e* and *f*, and the presence of *d* that of *f*. The distribution would be (as shown in Fig. 65):

| Quantity. Dollars. | Frequency. |
|---|---|
| 0 | 1 |
| 1 | 2 |
| 2 | 3 |
| 3 | 6 |
| 4 | 12 |
| 5 | 24 |
| 6 | 16 |

Suppose the presence of *a* or *b* or *c* implies the other two of the three, and that the presence of *e* implies that of *f*, and *vice versa*. The distribution will be (as shown in Fig. 66):

| Quantity. Dollars. | Frequency. |
| --- | --- |
| 0 | 1 |
| 1 | 1 |
| 2 | 3 |
| 3 | 10 |
| 4 | 7 |
| 5 | 19 |
| 6 | 23 |

It is clear then that the interdependence of the causes determining the quantity of a trait may cause all sorts of departures from the normal type of distribution, skewnesses and multimodal conditions, etc.; may, in less technical terms, cause the amounts of it appearing in an individual's different records or in the different individuals of a group to vary in all sorts of ways. In the illustration only simple and total dependencies were considered. Complex and partial dependencies would complicate the results to a well-nigh endless extent.

It should, however, be noted that if the causes are numerous and their interdependences of a random, hit-or-miss character, their combined action may be practically identical with that of totally independent causes. Thus, to continue with the same illustration, if there were five hundred relatives they might plan together in groups on various ways to give or withhold, and yet the final resultant, the probable total of John's Christmas income, might show no considerable differences from the total in case they had all acted independently.

A similar principle holds with reference to the equivalence of the causes in amount. In our illustration we demanded perfect equality, and a little experimentation will convince the reader that the approximation to the normal surface of frequency tends to disappear if $a > b > c > d$, etc.* However, with many causes and with a not too

* For instance, let the cause *a* equal 10, *b* equal 5 and *c*, *d*, *e* and *f* each equal 1. Then instead of the distribution of Table XXVI. we have (as shown in Fig. 67):

| Quantity. Dollars. | Frequency. | Quantity. Dollars. | Frequency. |
| --- | --- | --- | --- |
| 0 | 1 | 10 | 1 |
| 1 | 4 | 11 | 4 |
| 2 | 6 | 12 | 6 |
| 3 | 4 | 13 | 4 |
| 4 | 1 | 14 | 1 |
| 5 | 1 | 15 | 1 |
| 6 | 4 | 16 | 4 |
| 7 | 6 | 17 | 6 |
| 8 | 4 | 18 | 4 |
| 9 | 1 | 19 | 1 |

great variation in the amounts, the resulting distribution may closely mimic the perfectly normal type.

Finally, it should be remembered that the illustration taken is untrue to the common conditions of life in one respect. For these show us, not a group of causes, a chance combination from which determines the event, but such a group acting *together with some constant cause or set of causes.* Stature, strength, memory, wage-earning capacity, are due to certain constant causes which always act on all, plus a group, the action of which may be regarded in the mathematical fashion of this chapter. The addition of such a constant set of causes does not, of course, alter the form of distribution in the least, but simply adds the same amount to all its quantities, pushes them all ahead on the scale. In our illustration the $a$'s, $b$'s, $c$'s, etc., might more properly be the amounts which different friends might or might not give in addition to minimum sums, $k$, $k_1$, $k_2$, etc., which they always give, or be the gifts of some friends, who could not be counted on, superadded to a set of inevitable gifts $x$, $y$, $z$, etc., from a few.

The commoner method of describing the type of causation resulting in a normal surface of frequency of the amount of a trait starts with the presupposition that a certain amount tends to be and considers the causes as increasing or decreasing this. It is also common to use the frequencies, not of amounts of some continuous quantity, but of different proportions of black to white, or the like, in a chance draw of balls. The principles involved are precisely the same as those which have appeared in the more readily understood cases used here.

I have so far tried especially to show how the cooperation of a number of causes, each of which has a given likelihood of acting, may produce in the trait due to them a distribution of the so-called normal type. Incidentally, it has been noted that in general the form of distribution of any variable trait is due to the number of causes that influence its amount, their magnitude and their interrelations.

The form of distribution then is purely a secondary result of a trait's causation. There is no typical form or true form. There is nothing arbitrary or mysterious about variability which makes the normal type of distribution a necessity, or any more rational than any other sort or even any more to be expected on *a priori* grounds. Nature does not abhor irregular distributions.

On *a priori* grounds, indeed, the probability curve distribution would be exactly shown in any actual trait only by chance. For only by chance would the necessary conditions as to causation be fulfilled. And in point of fact, as the reader has constantly been told by the adjective 'approximate,' the exact probability curve distribution does not appear in the facts or give signs of being at the bottom of the facts of mental life. The common occurrence of distributions approaching it is due, not to any wonderful tendency of a group of cooperating causes to act so as to mimic the combinations of mathematical quantities equal and equally probable, but to the fact that many traits in human life are due to certain constant causes plus many occasional causes largely unrelated, small in amount in comparison with the constant causes and of the same order of magnitude among themselves.

It is the folly of the ignoramus in statistics to neglect the application of the algebraic laws of combinations to variable phenomena; it would be the folly of the pedant to try to bend all the variety of nature into conformity with one particular case of the frequency of combinations.*

The student interested in this subject should read some standard account of the algebra of combinations and probability, and Part II. of Bowley's 'Elements of Statistics.' Further references will be found on page 327 of the latter.

*It is a question whether students of mental measurement should not from the beginning be taught to put the so-called normal distribution in its proper place as simply one amongst an endless number of possible distributions, each and all due to and explainable by the nature of the causes determining the variations in the trait. The frequency of the occurrence of distributions somewhat like it could then be explained by a *vera causa*, the frequency of certain sorts of causation. On general principles this seems desirable, but in order to make for the student connections with the common discussions of statistical theory and practice and with the concrete work that has been done with mental measurements, I have compromised and subordinated the general *rationale* of the form of distribution to the explanation of the probability curve type.

# CHAPTER VI.

## THE ARITHMETIC OF CALCULATING CENTRAL TENDENCIES AND VARIABILITIES.

THE arithmetic of calculating averages, medians, modes, $\sigma$'s, A. D.'s, P. E.'s and other measures of central tendency and of variability is simple and straightforward if one bears in mind (1) that mental and social quantities are commonly continuous, so that any figure given as a measure means not a point, but a *distance* on the scale, and (2) that this distance is often that from the given figure to the next figure, so that the real value of the figure is itself plus one half of the unit of the scale.

The short methods of obtaining averages, $\sigma$'s and A. D.'s by guessing at the value and then correcting, are, however, foreign to the mathematical habits of one's school days and ordinarily require systematic practice before one gains surety and facility in their use. It will probably be advisable for the student to test himself with many simple problems, proving his result by the use of the longer traditional methods. In this and later mumerical work it will be of assistance to have at hand Crelle's '*Rechentafeln,*' which enable one to multiply and divide by numbers up to 1,000 with no labor save for eyes and fingers, and Barlow's '*Tables,*' which give the squares and square roots of all numbers up to 10,000.

The labor of calculating averages can be much reduced by adopting the method which most of us would probably use in a case like this: To get the average of 54, 52, 64, 56 and 50. Remembering that the average is such a figure that the sum of differences between it and the measures above it is equal to the sum of the differences between it and the measures below it, one takes 56 as the average. The differences below are 2, 4 and 6, that above is 8. If the average was altered by $-.8$, or to 55.2, the differences below would be 1.2, 3.2 and 5.2, and those above would be 8.8 and .8. This common procedure consists in guessing at an approximate average and then correcting it from knowledge of the sums of the minus and plus deviations from it. It lets us add small numbers instead of large and, as will be seen, gives us at the same time as the average, an approximate measure of the average deviation from it.

The choice of an approximate average is commonly easy after an inspection of the total distribution, and one soon acquires skill in making a correct choice in any case.

Suppose the measures to be as follows:

REACTION-TIMES OF V. H.

| Quantity. Seconds. | Frequency. |
|---|---|
| .120–124.99 or .1225 | 2 |
| .125 | 3 |
| .130 | 11 |
| .135 | 13 |
| .140 | 11 |
| .145 | 13 |
| .150 | 7 |
| .155 | 8 |
| .160 | 13 |
| .165 | 8 |
| .170 | 1 |
| .175 | 3 |
| .180 | 3 |
| .185 | 0 |
| .190 | 0 |
| .195 | 1 |

Either .145 — .1499 (*i. e.*, .1475) or .150 — .154 (*i. e.*, .1525) would do for a guess. I will use .145 — .1499. We have then to obtain the minus and plus deviations from .1475, the central point of the .145 — .1499 group. To save labor in multiplication and addition I shall measure these in terms, not of units of the scale, but of steps of the scale, *i. e.*, using five thousandths of a second as the unit. We have then for minus and plus deviations:

| | | |
|---|---|---|
| 2 | deviations of | — 5 or — 10 |
| 3 | " " | — 4 or — 12 |
| 11 | " " | — 3 or — 33 |
| 13 | " " | — 2 or — 26 |
| 11 | " " | — 1 or — 11 |
| 40 | | — 92 |

| | | |
|---|---|---|
| 7 | deviations of | + 1 or + 7 |
| 8 | " " | + 2 or + 16 |
| 13 | " " | + 3 or + 39 |
| 8 | " " | + 4 or + 32 |
| 1 | " " | + 5 or + 5 |
| 3 | " " | + 6 or + 18 |
| 3 | " " | + 7 or + 21 |
| 0 | | |
| 0 | | |
| 1 | " " | + 10 or + 10 |
| 44 | | + 148 |

The approximate average is evidently too low. It can be corrected by adding to it the algebraic sum of the deviations divided by the number of cases. In the illustration this will be $+\frac{56}{97}$ or $+.58$. .58 of a step = 2.9 thousanths of a second. The corrected average is then .1475 + .0029 or .1504 sec. Calling the algebraic sum of the deviations divided by the number of cases $d_{\text{act. av.}-\text{approx. av.}}$, we may summarize this whole calculation in the formulæ:

$$\text{Av.}_{\text{act.}} = \text{Av.}_{\text{approx.}} + d_{\text{act. av.}-\text{approx. av.}}$$

$$d_{\text{act. av.}-\text{approx. av.}} = \frac{\Sigma x\,(\text{algebraic})}{n}$$

*Determination of the Mode.*

In determining the mode one should seek not only the measure that is the most frequent on the basis of the limited series of measures he has before him, but also the one that would probably be the most frequent if a very great number of measures were at hand. There are two convenient tests of the latter fact. The mode from an infinite series of measures will probably be a measure representing the acme or culmination of a somewhat steady tendency of neighboring measures to greater and greater frequency. Graphically speaking it will be the apex of a slope. Hence we may consider the general tendency of the surface as a whole to rise to a maximum, grouping the cases so as to show a fairly regular rise, and use this knowledge in deciding the probable mode.

Doing this in the present case we get:

| Ability. | Frequency. | Ability. | Frequency. |
|---|---|---|---|
| 115–124.99 | 2 | 120–134.99 | 16 |
| 125 | 14 | 135 | 37 |
| 135 | 24 | 150 | 28 |
| 145 | 20 | 165 | 12 |
| 155 | 23 | 180 | 3 |
| 165 | 9 | 195 | 1 |
| 175 | 6 | | |
| 185 | 0 | | |
| 195 | 1 | | |

.145 up to .150 is probably the best choice for a mode.

The mode may be obtained from a quarter, then from a half, then from three quarters of the cases taken at random, and the influence of the increase in number of cases upon the position taken by the mode may be used to prophesy what position it would probably take with a very great number of cases.

Commonly with 200 or more measurements and with a grouping into not over 18 divisions, the mode is clear enough.*

The series of measures of Table XXVIII. may be taken as an example. 26 to 30 is the choice for a broad mode and 28 to 30 the best choice for a narrow one.

TABLE XXVIII.

MONEY AVAILABLE FOR SCHOOL PURPOSES DIVIDED BY AVERAGE ATTENDANCE; THAT IS, COST PER PUPIL FOR FULL YEAR'S ACTUAL ATTENDANCE. CITIES OF U. S. REPORT OF COM. OF ED., 1901.

| Quantity. Dollars. | Frequency. | Frequency in wider grouping. | Quantity. Dollars. | Frequency. | Frequency in wider grouping. |
|---|---|---|---|---|---|
| 10–11.99 | 6 | 11 | 4 | 5 | 9 |
| 12– | 5 | | 6 | 4 | |
| 14 | 10 | 24 | 8 | 6 | 9 |
| 6 | 14 | | 60 | 3 | |
| 8 | 20 | 36 | 2 | 2 | 6 |
| 20 | 16 | | 4 | 4 | |
| 2 | 31 | 60 | 6 | 5 | 9 |
| 4 | 29 | | 8 | 4 | |
| 6 | 34 | 73 | 70 | 2 | 2 |
| 8 | 39 | | 2 | 0 | |
| 30 | 31 | 61 | 4 | 2 | 2 |
| 2 | 30 | | 6 | 0 | |
| 4 | 24 | 42 | 8 | 2 | 3 |
| 6 | 18 | | 80 | 1 | |
| 8 | 17 | 39 | 2 | 2 | 2 |
| 40 | 22 | | 4 | 0 | |
| 2 | 16 | 32 | 6 | 0 | 0 |
| 4 | 16 | | 8 | 0 | |
| 6 | 15 | 29 | 90 | 1 | 2 |
| 8 | 14 | | 2 | 1 | |
| 50 | 3 | 13 | 4 | 0 | 1 |
| 2 | 10 | | 6 | 1 | |
| | | | | 465 | |

*Determination of the Median.* — The median is the $[(n+1)/2]^{th}$ measure.

Count in from each end, putting down occasionally the sums from the beginning. As the median is approached put them all

*These rough and ready methods of estimating the probably most frequent measure serve for any studies likely to be made by the non-mathematical student. A convenient account of a more precise method will be found in the *Journal of the Royal Statistical Society* for 1896, pp. 343–346.

down. The median will then fall among the cases of some one measure, $X$ (Case I.) or exactly between two measures, $X$ and $X_2$. In the latter case the measures may be side by side on the scale (Case II.) or separated by one or more measures the frequency of which is zero (Case III.). Case I. is, of course, by far the most common. Examples are given below:

| Quantity. | Case I. Frequency. | Case II. Frequency. | Case III. Frequency. |
|---|---|---|---|
| 9 up to 10 | 2 | 3 | 1 |
| 10 " 11 | 4 (6) | 5 (8) | 2 |
| 11 " 12 | 9 (15) | 11 (19) | 3 (6) |
| 12 " 13 | 14 (29) | 17 (36) | 8 (14) |
| 13 " 14 | 16 | 16 (36) | |
| 14 " 15 | 13 (27) | 12 (20) | 7 (14) |
| 15 " 16 | 8 (14) | 4 (8) | 1 (7) |
| 16 " 17 | 5 | 2 | 4 |
| 17 " 18 | 1 | 2 | 2 |
| | Median = 13.4 | Median = 13.0 | Median = 13.5 |

In Case I. take the percentage of the cases of the one measure in which the median lies needed to make the sum from the beginning one half the total number of cases; add this to the low limit of $X$ or subtract it from the upper limit of $X$, according to the direction in which you are taking the sums from the beginning, and the result is the median.* It is often a sufficiently close approximation to take simply the central value of $X$.

In Case II. take the upper limit of $X_1$ or the lower limit of $X_2$ which are of course the same thing.

In Case III. take the amount half-way between the upper limit of $X$ and the lower limit of $X_2$.

*Determination of the Average Deviation from the Average.*

The A. D. from the approximate average is the sum of the deviations of the individual measures from it (regardless of signs) divided by the number of cases. This sum is given in the course of the calculation of the average by our method. In the illustration it is 92 + 148, or 240.

A. D. from App. Av. = $\frac{240}{97}$, or 2.475. The step being 5, this

*This is not absolutely exact since the frequency of the different measures $X$ low limit + .1, $X$ low limit + .2, etc., will rarely be exactly the same, but it is sufficiently accurate for any mental measurements the student will encounter. A correction is possible only when the exact form of the distribution is known.

is, in thousandths of a second, 12.375; in seconds, .0124. This is incorrect (1) in that the 13 measures .145 to .150 have been regarded as all at 0 distance from .1475, whereas they would really deviate from it even up to half a step. Thus our A. D. is too small. On the other hand, (2) our figure is incorrect in that the measures of each group are regarded as centering at its mid-point, whereas really there would, as a rule, be more of them in the half of it nearer the average than in the other half. Thus our A. D. is too large. Corrections can be made for both of these errors, but in practice it does well enough to compute variabilities from a fine grouping, say into at least 15 groups, and then neglect the very small errors resulting, since they partially counterbalance each other.

Finally, the sum of the deviations from the actual averages will differ from the sum of those from an approximate average. It is easy to correct for this. In the illustration the 40 deviations below should each be increased .58 of a step, the 44 above each decreased .58 of a step, and the 13 zero deviations be changed each to —.58. This would give an increase of $9 \times .58$ step. This would alter the A. D. to .0123. This correction too may be neglected if the approximate average is chosen within one step. If it is not, it is often as easy to recalculate the deviations from the actual average, or a point very near it, as to make the correction.

These three errors may be called the errors of neglect of near deviations, of coarse grouping, and of the approximate average.

*Determination of the Standard Deviation from the Average.*

Obtain the sum of the square of the deviations from the approximate average or, if it is not within one step of the actual average, of the deviations from a point that is. Then calculate $\sigma$ from the formula $\sqrt{(\Sigma x^2)/n}$, the $x$'s equaling the deviations from the point chosen. The corrections for the errors of neglect of near deviations, of coarse grouping, and of the approximate average may be left uncorrected without serious inaccuracy, as in the case of the A. D. The correction for the last is to subtract $d^2$, $d$ equaling, as before, $(\Sigma x)/n$ (algebraic).

In the illustration if 150, that is, a point just between the 145–150 and 150–155 groups, is taken as the point from which to get an approximate $\sigma$, the calculation is as follows:

$$
\begin{aligned}
2 \times (5.5)^2 &= 60.50 \\
3 \times (4.5)^2 &= 60.75 \\
11 \times (3.5)^2 &= 134.75 \\
13 \times (2.5)^2 &= 81.25 \\
11 \times (1.5)^2 &= 24.75 \\
13 \times (\ .5)^2 &= 3.25 \\
\hline
&\phantom{=}\ 365.25
\end{aligned}
$$

$$
\begin{aligned}
7 \times (\ .5)^2 &= 1.75 \\
8 \times (1.5)^2 &= 18.00 \\
13 \times (2.5)^2 &= 81.26 \\
8 \times (3.5)^2 &= 98. \\
1 \times (4.5)^2 &= 20.25 \\
3 \times (5.5)^2 &= 90.75 \\
3 \times (6.5)^2 &= 126.75 \\
0 \times (7.5)^2 &= \\
0 \times (8.5)^2 &= \\
1 \times (9.5)^2 &= 90.25 \\
\hline
&\phantom{=}\ 527.00
\end{aligned}
$$

$$365 + 527 = 892$$

$$\sqrt{\tfrac{892}{97}} = \sqrt{9.2}$$

$\sqrt{9.2} = 3.033$ or, in seconds, .01516. $\sigma = .01516$.

It is much easier to take as an arbitrary step one half the regular step in cases where the chosen point is just between two groups. We then have whole numbers to deal with. The above would become :

$$
\begin{aligned}
2 \times (11)^2 &= 242 \\
3 \times (\ 9\ )^2 &= 243 \\
11 \times (\ 7\ )^2 &= 539 \\
13 \times (\ 5\ )^2 &= 275 \\
11 \times (\ 3\ )^2 &= 99 \\
13 \times (\ 1\ )^2 &= 13 \\
\hline
&\phantom{=}\ 1411
\end{aligned}
$$

$$
\begin{aligned}
7 \times (\ 1\ )^2 &= 7 \\
8 \times (\ 3\ )^2 &= 72 \\
13 \times (\ 5\ )^2 &= 275 \\
8 \times (\ 7\ )^2 &= 392 \\
1 \times (\ 9\ )^2 &= 81 \\
3 \times (11)^2 &= 363 \\
3 \times (13)^2 &= 507 \\
0 \times (15)^2 &= \\
0 \times (17)^2 &= \\
1 \times (19)^2 &= 361 \\
\hline
&\phantom{=}\ 2058
\end{aligned}
$$

$$1411 + 2058 = 3569$$

$$\sqrt{\tfrac{3569}{97}} = \sqrt{36.8}$$

$\sqrt{36.8} = 6.07$, or the step in this case being $\frac{5}{2}$ instead of 5 as before, .01518 sec.

The object of calculating the variability from an approximate average is, of course, to save the multiplication, addition and squaring of long numbers. In general, it may be said of mental, social and physiological measurements that it is wise to save labor in their calculation so as to expend it in getting more or more accurate measurements. By the methods given here calculations can be made very rapidly.

*The Determination of the P. E. from the Average.*

The P. E. equals the amount of deviation from the average (regardless of signs) which is exceeded by exactly 50 per cent. of the deviations of the individual measures. To obtain it directly, arrange these deviations in the order of magnitude and find the point reached in counting off half of them. For instance, in the case on page 72 the deviations from .1504 are:

| | | |
|---|---|---|
| Between 0 and .0004 in one direction and between | 0 and .0046 in the other | 7 |
| " .0004 and .0054 " " " " | .0046–.0096 | 21 |
| " .0054 and .0104 " " " " | .0096–.0146 | 24 |
| " .0104 and .0154 " " " " | | 13 |

The total number of cases being 97, it is sure that the P. E. is somewhere between .0054 and .0154.

If the measurements were on a finer scale, it could be located more accurately and still be sure.

*A*. So also if the average fell exactly at the mid-point of a group or just between two groups. For instance, if the average in the present case were .150, the deviations would rank

| | |
|---|---|
| Between 0 and .005 | 20 |
| " .005 and .010 | 19 |
| " .010 and .015 | 26 |

The P. E. would then surely be between .010 and .015. We could also assume that the $9\frac{1}{2}$ of the 26 deviations between .010 and .015, which are needed to bring us to the 50 per cent. point, will bring us approximately 9.5/26 of the distance* from .010 to

* Really a little less, because of the greater frequency of measures near the average than of those more remote from it within the groups .135–.140 and .160–.165.

.015, that is, to .0118. The P. E. then would be approximately .0118.

*B.* In so far as the measurements are distributed symmetrically about the average, the P. E. calculated directly will be the same as the distance from the average reached by counting off in either direction 25 per cent. of $N$ (the total number of measures in the distribution). This would again be the same as the distance from the average reached by counting in 25 per cent. of $N$ from either extreme.

*A* and *B* give two ways of reaching quickly an approximate P. E. The P. E. calculated from the mid-point nearest the average or from the point between two groups nearest the average will be a close approximation to the P. E. from the actual average. Its calculation as in *A* is easy.

In so far as the distribution is approximately symmetrical (and when it is not, any single measure of the variability should be replaced by two — one of the variability above, the other of the variability below), half the distance between the 25 percentile and 75 percentile gives a very close approximation to the P. E.

*Determination of Quartiles, Octiles and Other Percentile Values.*

The determination of these measures has only one difficulty, that of allowing for the form of the distribution, which commonly makes cases within any group more frequent near the average. For instance, if we wish to find the lower octile in the case given on page 79, we have $n = 465$, $\frac{1}{8}n = 58.125$, and up to measure 20, 55 cases, 3.125 cases more will bring us to the octile point. How far will they bring us from 20 toward 22. If the 16 cases above 20 and below 22 were evenly distributed, if 20.1, 20.2, 20.3, etc., were equally frequent, it would be correct to take 3.125/20 of 2 as the distance above 20 to be traversed. But the general form of the distribution tells us that the measures near the mode are more likely to occur. For perfect exactness an allowance should be made. If the groups into which the distribution is divided are few in number this allowance is of some importance, but when the division is into 15 or more groups, the simple percentage method will be sufficiently exact to determine quartiles and exact enough to determine octiles for any use to which they will probably be put.

*Determination of the Average Deviation and of the Standard Deviation from the Median.*

The method is identical with that described under 'Determination of A. D. and of $\sigma$ from the Average,' except that the approximate average there should be replaced by 'approximate median' and that the $d$ (Act. Av. and App. Av.) should be replaced by $d$ (Act. Median and App. Median). The $d$ will here be calculated directly.

*Determination of the P. E. from the Median.*

The P. E. may be obtained directly, but for approximately symmetrical distributions the $B$ method on page 79 is accurate enough and much quicker, *viz.*, count in from the low end until 25 per cent. of the cases * are covered. Call the quantity thus reached the 25 percentile. Do likewise from the high end to obtain the 75 percentile. P. E. = approximately $\frac{1}{2}$ (75 percentile − 25 percentile).

It is wise, in general, to also present the values *75 percentile − median* and *median − 25 percentile,* which represent the variability below separately from that above the median. If there is a constant difference between the two in series of measures of any one sort, both should be given to show the skewness of distribution.

*Determination of Various Percentile Values.*

The limits about the median needed to include any given percentage of cases can be found in the same way.

*Determination of the Average Deviation and Standard Deviation from the Mode.*

The method is identical with that described under 'Determination of A. D. and of $\sigma$ from the average,' except that the 'Approximate Average' should be replaced by mode and that no correction is needed, the formulæ being simply :

$$\text{A. D. from mode} = \Sigma x / n,$$

$$\sigma \text{ from mode} = \sqrt{\Sigma x^2 / n}.$$

* If the sums from the beginning have been jotted down during the calculation of the median, the 25 and 75 percentile points can be found in less than a minute.

### *Determination of the P. E. from the Mode.*

The P. E. may be calculated directly with little labor, if an integral measure is taken as the mode. In other cases follow the *A* method of approximation.

### *Determination of Various Percentile Values from the Mode.*

The methods already given suffice.

When variabilities are measured from the average of a skewed distribution (the mode should, in the majority of skewed distributions, be used instead) the variability above and that below the average should be given separately. That is, the distribution should be divided into the cases above and the cases below the central tendency, *c*. Call these $n_a$ and $n_b$. Then find the average deviation of the $n_a$ group from *c* and also the average deviation of the $n_b$ group from *c*. For $\sigma$ do the same. Instead of the P. E. get such values as, half the cases of $n_a$ deviate less than so much from *c*; half the cases of $n_b$ deviate less than so much from *c*; one fourth of the cases of $n_a$ deviate less than so much from *c*, etc. The methods of approximation allowed hitherto may be used. A sample calculation is given below.

In multimodal distributions the variability should be calculated separately for the distributions into which the given distribution should be analyzed.

| Quantity. | Frequency. | Sums from Beginning. |
|---|---|---|
| 21, *i. e.*, 20.5 to 21.52 | 2 | |
| 22 | 5 | 7 |
| 23 | 16 | 23 |
| 24 | 40 | 63 |
| 25 | 60 | 123 |
| 26 | 92 | 215 |
| 27 | 100 | 315 |
| 28 | 120 | |
| 29 | 96 | 531 |
| 30 | 84 | 435 |
| 31 | 80 | 351 |
| 32 | 70 | 271 |
| 33 | 62 | 201 |
| 34 | 48 | 139 |
| 35 | 36 | 91 |
| 36 | 20 | 55 |
| 37 | 14 | 35 |
| 38 | 10 | 21 |
| 39 | 6 | 11 |
| 40 | 2 | 5 |
| 41 | 2 | 3 |
| 42 | 1 | |

$$n = 315 + 120 + 531, \; i.\,e., \; n = 966,$$

$$\text{mode} = 28, \; i.\,e., \; 27.5 \text{ to } 28.5,$$

$$n_a = 531 + 60, \; i.\,e., \; n_a = 591, \quad n_b = 315 + 60, \; i.\,e., \; n_b = 375,$$

$$\tfrac{1}{2}n_a = 295.5, \quad \tfrac{1}{2}n_b = 187.5.$$

Points reached in counting in 295.5 from 42 and 187.5 from 21 are

$$31.5 - [(24.5/80) \times 1] \text{ and } 25.5 + [(64.5/92) \times 1].$$

These are 31.2 and 26.2.

$\frac{1}{2}n_a$ are less than 3.2 distant from the mode.
$\frac{1}{2}n_b$ " " " 1.8 " " "

## Problems.

16. Calculate the average and the A. D. and $\sigma$ from it in each of the following cases; also the median and 25 and 75 percentiles. Obtain results accurate within .5 the unit.

Case I.

| Quantity. | Frequency. |
|---|---|
| 11.00 | 2 |
| 12.00 | 1 |
| 13.00 | 4 |
| 14.00 | 9 |
| 15.00 | 21 |
| 16.00 | 11 |
| 17.00 | 6 |
| 18.00 | 1 |
| 19.00 | 1 |

Case II.

| Quantity. | Frequency. |
|---|---|
| 140 up to 144 | 1 |
| 144 " 148 | 1 |
| 148 " 152 | 4 |
| 152 " 156 | 7 |
| 156 " 160 | 13 |
| 160 " 164 | 20 |
| 164 " 168 | 22 |
| 168 " 172 | 15 |
| 172 " 176 | 5 |
| 176 " 180 | 2 |
| 180 " 184 | 2 |

Case III.

| Quantity. | Frequency. |
|---|---|
| 3 up to 4 | 1 |
| 4 " 5 | 3 |
| 5 " 6 | 1 |
| 6 " 7 | 3 |
| 7 " 8 | 4 |
| 8 " 9 | 4 |
| 9 " 10 | 10 |
| 10 " 11 | 13 |
| 11 " 12 | 13 |
| 12 " 13 | 18 |
| 13 " 14 | 16 |
| 14 " 15 | 9 |
| 15 " 16 | 15 |
| 16 " 17 | 20 |
| 17 " 18 | 10 |
| 18 " 19 | 6 |
| 19 " 20 | 7 |
| 20 " 21 | 3 |
| 21 " 22 | 1 |
| 22 " 23 | 2 |
| 23 " 24 | 2 |
| 24 " 25 | 2 |
| 25 " 26 | 0 |
| 26 " 27 | 2 |

17. In each of the following cases, calculate the average and the A. D., $\sigma$ and P. E. from it; the median and the A. D., $\sigma$ and P. E from it. Accuracy to .5 the unit.

| Case I. Number of *A*'s marked. | Case I. Frequency. | Case II. Temperature at mouth. | Case II. Frequency. |
|---|---|---|---|
| 14 up to 16 | 2 | 96.0 up to 96.2 | 2 |
| 16, etc. | 0 | 96.2, etc. | 0 |
| 18 | 2 | 96.4 | 0 |
| 20 | 2 | 96.6 | 0 |
| 22 | 6 | 96.8 | 0 |
| 24 | 3 | 97.0 | 3 |
| 26 | 10 | 97.2 | 2 |
| 28 | 12 | 97.4 | 3 |
| 30 | 17 | 97.6 | 4 |
| 32 | 28 | 97.8 | 3 |
| 34 | 16 | 98.0 | 25 |
| 36 | 30 | 98.2 | 20 |
| 38 | 25 | 98.4 | 13 |
| 40 | 30 | 98.6 | 28 |
| 42 | 22 | 98.8 | 14 |
| 44 | 23 | 99.0 | 15 |
| 46 | 23 | 99.2 | 4 |
| 48 | 13 | 99.4 | 7 |
| 50 | 11 | 99.6 | 7 |
| 52 | 11 | 99.8 | 1 |
| 54 | 11 | 100.0 | 1 |
| 56 | 2 | 100.2 | 1 |
| 58 | 1 | 100.4 | 1 |
| 60 | 4 | | |
| 62 | 5 | | |
| 64 | 0 | | |
| 66 | 1 | | |
| 68 | 0 | | |
| 70 | 1 | | |
| 72 | 0 | | |
| 74 | 0 | | |
| 76 | 0 | | |
| 78 | 1 | | |

18. In Case II. of 17, what reasons are there for supposing that the grouping that follows is truer to the real facts than are the actual reported measures? Calculate average and A. D. for this second grouping.

| | |
|---|---|
| 96.2 up to 96.6 | 1 |
| 96.6, etc. | 1 |
| 97.0 | 5 |
| 97.4 | 7 |
| 97.8 | 28 |
| 98.2 | 33 |
| 98.6 | 42 |
| 99.0 | 19 |
| 99.4 | 14 |
| 99.8 | 2 |
| 100.2 | 2 |

19. In each of the following cases, determine the mode and the variability of the distribution around it. Calculate also the average and the variability around it.

CASE I.

| Weight of Adult Englishmen. * | Frequency. |
|---|---|
| 90 lbs. up to 100 lbs. | 2 |
| 100 etc. | 26 |
| 110 | 133 |
| 120 | 338 |
| 130 | 694 |
| 140 | 1240 |
| 150 | 1075 |
| 160 | 881 |
| 170 | 492 |
| 180 | 304 |
| 190 | 174 |
| 200 | 75 |
| 210 | 62 |
| 220 | 33 |
| 230 | 10 |
| 240 | 9 |
| 250 | 3 |
| 260 | 1 |

CASE II.

| Different Rates of Interest. | Quantity (of money loaned). |
|---|---|
| 4.00 | 1014 |
| 4.25 | 45 |
| 4.375 | 40 |
| 4.50 | 17232 |
| 4.75 | 1711 |
| 5.00 | 22987 |
| 5.17 | 21 |
| 5.25 | 242 |
| 5.50 | 3293 |
| 5.75 | 27 |
| 6.00 | 6955 |
| 6.25 | 52 |
| 6.50 | 158 |
| 6.67 | 1449 |
| 6.75 | 59 |
| 7.00 | 2263 |
| 7.25 | 7 |
| 7.50 | 306 |
| 8.00 | 1585 |
| 8.50 | 892 |

20. In the report † from which Case II. is quoted the 4.0 = 4 or less and the 8.5 = 8.5 or more. If these facts had been announced in the problem, which measures only could have been calculated?

* Roberts' 'Manual of Anthropometry' is the source of these figures.

† New Zealand Official Year-Book, 1901, p. 231.

# CHAPTER VII.

## THE TRANSMUTATION OF MEASURES BY RELATIVE POSITION INTO TERMS OF UNITS OF AMOUNT.

IF a group of individuals are ranged in order according to the amounts which they severally possess of a trait, we can, even when ignorant of what the amounts are for each and all of the individuals, assign to each the amount of his deviation from the average, provided the form of the group's distribution is known.

For instance, let 100 boys rank with respect to scholarship as shown in Table XXIX., and let the form of distribution be that of Fig. 68.

TABLE XXIX.

100 BOYS *a*, *b*, *c*, ETC., RANKED BY RELATIVE POSITION.

| | | |
|---|---|---|
| 1 | a is the highest ranking boy. | |
| 3 | b, c, d are the next highest ranking and are indistinguishable. | |
| 6 | e, f, g, h, i, j | " |
| 10 | k, l, m, n, o, p, q, r, s, t | " |
| 15 | u, v, w, x, y, z, *a*, *b*, *c*, *d*, *e*, *f*, *g*, *h*, *i* | " |
| 17 | *j*, *k*, *l*, *m*, *n*, *o*, *p*, *q*, *r*, *s*, *t*, *u*, *v*, *w*, *x*, *y*, *z* | " |
| 19 | A, B, C, D, E, F, G, H, I, J, K, L, M, N, O, P, Q, R, S | " |
| 14 | T, U, V, W, X, Y, Z, α, β, γ, δ, ε, ς, η | " |
| 8 | θ, ι, κ, λ, μ, ν, ξ, ο | " |
| 4 | π, ρ, σ, τ, | " |
| 3 | υ, φ, χ | " |

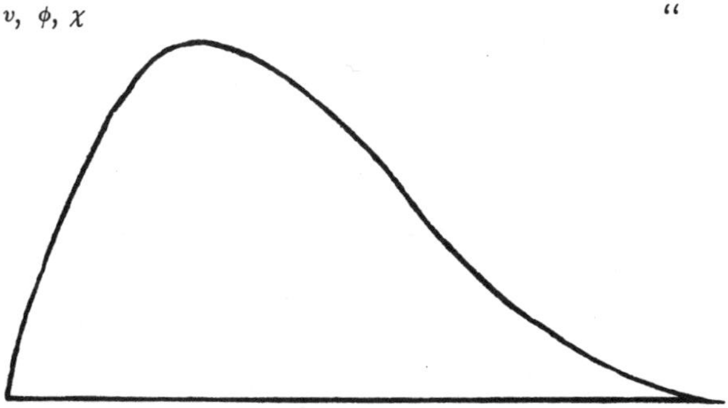

FIG. 68.

If we build up approximately the distribution of Fig. 68 by a series of 40 rectangles of equal base, the result is Fig. 69. Call the low extreme *A* and the length of base of each of the rectangles

$K$. Then the upper extreme is at $A + 40K$. The approximate distribution in terms of these units is given in Table XXX. The frequencies may, of course, be reckoned on the basis of any arbitrary unit. In Table XXX., the total area is taken to be 1,680.

TABLE XXX.

| Quantity. | Frequency. | Quantity. | Frequency. | Quantity. | Frequency. |
|---|---|---|---|---|---|
| $A$ to $A+K$ | 7 | $A+14K$ | 74.5 | $A+27K$ | 26 |
| $A+K$ to $A+2K$ | 20.5 | " 15 | 72.5 | " 28 | 22.5 |
| $A+$ $2K$, etc. | 23 | " 16 | 70 | " 29 | 19.5 |
| " $3K$ | 44 | " 17 | 66.5 | " 30 | 16.5 |
| " 4 | 52.5 | " 18 | 63.5 | " 31 | 14 |
| " 5 | 60.5 | " 19 | 60 | " 32 | 11.5 |
| " 6 | 67.5 | " 20 | 56 | " 33 | 9.5 |
| " 7 | 73.5 | " 21 | 52 | " 34 | 7.5 |
| " 8 | 77.5 | " 22 | 47.5 | " 35 | 5.5 |
| " 9 | 80 | " 23 | 42 | " 36 | 4 |
| " 10 | 80 | " 24 | 38 | " 37 | 2.5 |
| " 11 | 79.5 | " 25 | 34 | " 38 | 2 |
| " 12 | 78.5 | " 26 | 30 | $A+39K$ to $A+40K$ | .5 |
| " 13 | 77 | | | | |

The highest ranking boy, a, who was the top 1 per cent. of the group, will in our figure occupy the top 1 per cent. in the table, the highest 16.8 of the frequencies. His ability then is from $A + 40K$ part way into $A + 34K$. The abilities of the next three, b, c and

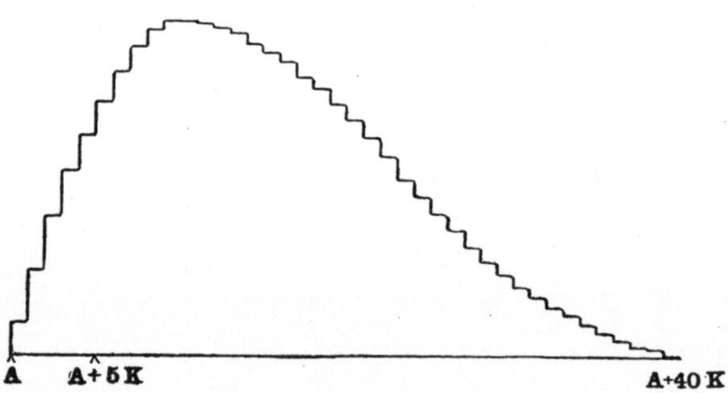


Fig. 69.

d, will occupy the next 50.4 of the frequencies and be included between the limits $A + 34.7K$ and $A + 30.4K$. So on with the next six and the rest. The limits for each group are shown in Fig. 70.

The average ability of each group may be calculated roughly *

* By a subdivision of the surface into finer rectangles the precision of these averages could have been increased.

from the facts obtained in this way. Thus the highest boy, being represented by 0.5 $(A + 39K)$, 2 $(A + 38.5K)$, 2.5 $(A + 37.5K)$, 4 $(A + 36.5K)$, 5.5 $(A + 35.5K)$ and 2.3 $(A + 34.5K)$, has as an average $A + 36.5K$.

A table can thus be formed as follows :

Boy a has as his ability $A + 36.5K$;
Boys b, c, d, have as their ability $A + 32.2K$;
Boys e, f, g, h, i, j, have as their ability $A + 28.0K$;
Boys k, l, m, n, o, p, q, r, s, t, have as their ability $A + 23.8K$; etc.

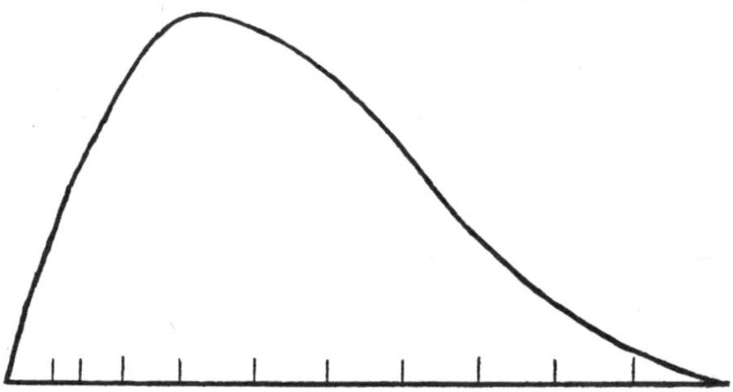

FIG. 70.

These measures can further be turned into distances from the mode or median or average of the distribution instead of from its lower limit $A$. They can be put in terms of any measure of the variability of the scheme, or of any part of it instead of $K$. For the distribution given in Table XXX. can be used in every way like one with known quantities in place of the $A$ and $K$. For instance, the best boy is $+ 26K$ from the mode, or, in units of the *75 percentile — mode* measure of variability, is + 3.38.

The scholarship of every boy in the group is thus represented in definite quantities of some unit of amount of difference from some standard. This unit itself is definable as the difference between this person and that person. The standard is similarly definable as the scholarship of such and such a person.

By this method the obscurest and most complex traits, such as morality, enthusiasm, eminence, efficiency, courage, legal ability, inventiveness, etc., can be made material for ordinary statistical procedure, the one condition being that the general form of distribution of the trait in question be approximately known.

If now one has a group of individuals ranked by their relative position in the group, his first task before he can transmute the series of relative positions into a series in units of amount is to ascertain the form of distribution. This may be done (1) by measuring objectively in units of amount enough sample individuals, or (2) if the trait cannot be measured in units of amount, by inferring the form of distribution from that of similar traits which can be.

1. Suppose one had 2,000 ten-year-old boys measured with respect to intellect by relative position.* If now one measured 200 of them objectively with tests scorable in units of amount, he could properly transmute the 2,000 on the basis of the type of distribution of the 200.

2. Suppose one had 1,000 individuals measured with respect to delicacy of discrimination of sound by relative position. (It is well-nigh impossible to measure sensitiveness to sound in objective units which another observer can duplicate, because of the influence of size of room, resonance, etc.) It is fairly certain from studies of the delicacy of discrimination of length, weight, etc., that delicacy of discrimination of sound is distributed in something approximating sufficiently to a probability surface, with range of from $+ 3\sigma$ to $- 3\sigma$, to prevent calculations on that basis from being more than a little wrong on the average. We may, therefore, transmute the 1,000 measures by relative position into units of amount, on the hypothesis that such is the form of distribution. So also with school marks if intellect in general is found to follow the probability type of distribution.

The labor of transmutation for cases which follow the probability type of distribution is rendered almost *nil* by the use of tables.

If the probability surface of range $+ 3\sigma$ to $- 3\sigma$ is divided up into 100 equal areas representing the 100 successive per cents. from the highest to the lowest of the total group, and the average distance from the average in terms of $\sigma$ is calculated for each per cent., the result is Table XXXI.

If now we ask, 'What will be the average ability of the highest 6 per cent.?' we have only to add the figures for the first 6 per cents. and divide by 6 (the result being, of course, 1.99). Similarly to get

* Such measures, at least approximately correct, would in fact be easy to obtain through school marks, teachers' opinions, personal conferences, etc.

TABLE XXXI.

VALUES, IN TERMS OF THE STANDARD DEVIATION $\sigma$, OF EACH SINGLE PER CENT., THE DISTRIBUTION BEING NORMAL. BEGINNING WITH THE EXTREME.

| Per cents. in order from highest value to mode or from lowest value to mode. | Value in terms of $\sigma$. | Per cents. in order from highest value to mode or from lowest value to mode. | Value in terms of $\sigma$. |
|---|---|---|---|
| 1st | 2.7 | 26th | .659 |
| 2d | 2.18 | 27th | .628 |
| 3d | 1.96 | 28th | .598 |
| 4th | 1.81 | 29th | .568 |
| 5th | 1.695 | 30th | .539 |
| 6th | 1.598 | 31st | .510 |
| 7th | 1.514 | 32d | .482 |
| 8th | 1.439 | 33d | .454 |
| 9th | 1.372 | 34th | .426 |
| 10th | 1.311 | 35th | .399 |
| 11th | 1.250 | 36th | .372 |
| 12th | 1.200 | 37th | .345 |
| 13th | 1.150 | 38th | .319 |
| 14th | 1.103 | 39th | .293 |
| 15th | 1.058 | 40th | .266 |
| 16th | 1.015 | 41st | .240 |
| 17th | .974 | 42d | .210 |
| 18th | .935 | 43d | .189 |
| 19th | .896 | 44th | .164 |
| 20th | .860 | 45th | .139 |
| 21st | .824 | 46th | .113 |
| 22d | .789 | 47th | .087 |
| 23d | .755 | 48th | .063 |
| 24th | .722 | 49th | .037 |
| 25th | .690 | 50th | .013 |

the average ability of any consecutive series of per cents. Table XXXII. gives the results of such computation for every consecutive series in the upper half of the total group. If the signs are changed to minus it serves for the lower half.

The figures along the top stand each for the per cent. already made up in counting in from the extremes. The figures down the side stand for the per cent. in the group for which a measure in terms of amount is to be found. The entries in the body of the table stand for the average amount, in terms of $\sigma$, of any per cent. counted in from any point to the average. When any per cent. passes the average (*e. g.*, 30 per cent., often 40 per cent., have been used up in counting in from the top) it is necessary to take from the table two entries, one for the plus cases down to the average, the other for the minus cases, up to the average, of which the per cent. is made up, and from these two entries to compute the average for the given per cent. Thus, 40 per cent. from the upper extreme having been used up, the next 30 per cent. will average

$$\frac{(+.13 \times 10) + (-.26 \times 20)}{30}, \text{ or } -.13.$$

Illustrations of the simpler usage in cases not passing the average are as follows:

| | |
|---|---|
| The first 1 per cent. of a group averages | + 2.7 |
| The " 8 " " " " average | + 1.86 |
| The 9th and 10th " " " " | + 1.34 |
| Per cents. 6, 7 and 8 from the bottom " | — 1.57. |

TABLE XXXII (*a*).

| | 0 | 1 | 2 | 3 | 4 | 5 | 6 | 7 |
|---|---|---|---|---|---|---|---|---|
| 1 | 270 | 218 | 196 | 181 | 170 | 160 | 151 | 144 |
| 2 | 244 | 207 | 189 | 175 | 165 | 156 | 148 | 141 |
| 3 | 228 | 198 | 182 | 170 | 160 | 152 | 144 | 137 |
| 4 | 216 | 191 | 177 | 165 | 156 | 148 | 141 | 134 |
| 5 | 210 | 185 | 172 | 161 | 152 | 145 | 138 | 131 |
| 6 | 199 | 179 | 167 | 157 | 149 | 141 | 135 | 129 |
| 7 | 192 | 174 | 163 | 153 | 145 | 138 | 132 | 126 |
| 8 | 186 | 170 | 159 | 150 | 142 | 135 | 128 | 124 |
| 9 | 181 | 165 | 155 | 147 | 139 | 133 | 126 | 121 |
| 10 | 176 | 161 | 151 | 143 | 136 | 130 | 124 | 119 |
| 11 | 171 | 158 | 148 | 140 | 134 | 127 | 122 | 116 |
| 12 | 167 | 154 | 145 | 138 | 131 | 125 | 119 | 114 |
| 13 | 163 | 151 | 142 | 135 | 128 | 122 | 117 | 112 |
| 14 | 159 | 147 | 139 | 132 | 126 | 120 | 115 | 110 |
| 15 | 156 | 144 | 136 | 129 | 123 | 118 | 113 | 108 |
| 16 | 152 | 141 | 134 | 127 | 121 | 116 | 111 | 106 |
| 17 | 149 | 139 | 131 | 125 | 119 | 113 | 109 | 104 |
| 18 | 146 | 136 | 129 | 122 | 117 | 111 | 106 | 102 |
| 19 | 143 | 133 | 126 | 120 | 114 | 109 | 105 | 100 |
| 20 | 140 | 131 | 124 | 118 | 112 | 107 | 103 | 98 |
| 21 | 137 | 128 | 121 | 116 | 110 | 105 | 101 | 96 |
| 22 | 135 | 126 | 119 | 113 | 108 | 103 | 99 | 95 |
| 23 | 132 | 124 | 117 | 111 | 106 | 101 | 97 | 92 |
| 24 | 130 | 121 | 115 | 109 | 104 | 100 | 95 | 91 |
| 25 | 127 | 119 | 113 | 107 | 102 | 98 | 93 | 89 |
| 26 | 125 | 117 | 111 | 105 | 101 | 96 | 92 | 88 |
| 27 | 123 | 115 | 109 | 104 | 99 | 94 | 90 | 86 |
| 28 | 120 | 113 | 107 | 102 | 97 | 92 | 88 | 84 |
| 29 | 118 | 111 | 105 | 100 | 95 | 91 | 87 | 83 |
| 30 | 116 | 109 | 103 | 98 | 93 | 89 | 85 | 81 |
| 31 | 114 | 107 | 101 | 96 | 92 | 87 | 83 | 79 |
| 32 | 112 | 105 | 99 | 94 | 90 | 86 | 82 | 78 |
| 33 | 110 | 103 | 98 | 93 | 88 | 84 | 80 | 76 |
| 34 | 108 | 101 | 96 | 91 | 86 | 82 | 79 | 75 |
| 35 | 106 | 99 | 94 | 89 | 85 | 81 | 77 | 73 |
| 36 | 104 | 97 | 92 | 88 | 82 | 80 | 75 | 72 |
| 37 | 102 | 96 | 91 | 86 | 82 | 78 | 74 | 70 |
| 38 | 100 | 94 | 89 | 84 | 80 | 76 | 72 | 69 |
| 39 | 98 | 92 | 87 | 83 | 79 | 75 | 71 | 67 |
| 40 | 97 | 91 | 86 | 81 | 77 | 73 | 69 | 66 |
| 41 | 95 | 89 | 84 | 80 | 75 | 72 | 68 | 64 |
| 42 | 93 | 87 | 82 | 78 | 74 | 70 | 66 | 63 |
| 43 | 91 | 85 | 81 | 76 | 72 | 69 | 65 | 62 |
| 44 | 90 | 84 | 79 | 75 | 71 | 67 | 64 | |
| 45 | 88 | 82 | 78 | 73 | 69 | 66 | | |
| 46 | 86 | 81 | 76 | 72 | 68 | | | |
| 47 | 85 | 79 | 75 | 70 | | | | |
| 48 | 83 | 78 | 73 | | | | | |
| 49 | 81 | 76 | | | | | | |
| 50 | 80 | | | | | | | |

TABLE XXXII (*b*).

| | 8 | 9 | 10 | 11 | 12 | 13 | 14 | 15 |
|---|---|---|---|---|---|---|---|---|
| 1 | 137 | 131 | 125 | 120 | 115 | 110 | 106 | 102 |
| 2 | 134 | 128 | 122 | 118 | 112 | 108 | 104 | 99 |
| 3 | 131 | 125 | 120 | 115 | 110 | 107 | 102 | 97 |
| 4 | 128 | 123 | 118 | 113 | 108 | 104 | 100 | 96 |
| 5 | 126 | 120 | 115 | 111 | 106 | 102 | 98 | 94 |
| 6 | 123 | 118 | 113 | 108 | 104 | 100 | 96 | 92 |
| 7 | 121 | 116 | 111 | 106 | 102 | 98 | 94 | 90 |
| 8 | 118 | 113 | 109 | 104 | 100 | 96 | 92 | 88 |
| 9 | 116 | 111 | 106 | 102 | 98 | 94 | 90 | 86 |
| 10 | 114 | 109 | 104 | 100 | 96 | 92 | 88 | 85 |
| 11 | 111 | 107 | 102 | 98 | 94 | 90 | 87 | 83 |
| 12 | 109 | 105 | 100 | 96 | 92 | 89 | 85 | 81 |
| 13 | 107 | 103 | 99 | 94 | 91 | 87 | 83 | 80 |
| 14 | 105 | 101 | 97 | 93 | 89 | 85 | 81 | 78 |
| 15 | 103 | 99 | 95 | 91 | 87 | 83 | 80 | 76 |
| 16 | 101 | 97 | 93 | 89 | 85 | 82 | 78 | 75 |
| 17 | 99 | 95 | 91 | 87 | 84 | 80 | 77 | 73 |
| 18 | 98 | 93 | 89 | 86 | 82 | 78 | 75 | 72 |
| 19 | 96 | 92 | 88 | 84 | 80 | 77 | 73 | 70 |
| 20 | 94 | 90 | 86 | 82 | 79 | 75 | 72 | 69 |
| 21 | 92 | 88 | 84 | 81 | 77 | 74 | 70 | 67 |
| 22 | 90 | 87 | 83 | 79 | 76 | 72 | 69 | 66 |
| 23 | 89 | 85 | 81 | 78 | 74 | 71 | 67 | 64 |
| 24 | 87 | 83 | 80 | 76 | 73 | 69 | 66 | 63 |
| 25 | 85 | 82 | 78 | 74 | 71 | 68 | 64 | 61 |
| 26 | 84 | 80 | 76 | 73 | 70 | 66 | 63 | 60 |
| 27 | 82 | 78 | 75 | 71 | 68 | 65 | 62 | 58 |
| 28 | 80 | 77 | 73 | 70 | 67 | 63 | 60 | 57 |
| 29 | 79 | 75 | 72 | 68 | 65 | 62 | 59 | 56 |
| 30 | 77 | 74 | 70 | 67 | 64 | 60 | 57 | 54 |
| 31 | 76 | 72 | 69 | 65 | 62 | 59 | 56 | 53 |
| 32 | 74 | 71 | 67 | 64 | 61 | 58 | 54 | 51 |
| 33 | 73 | 69 | 66 | 63 | 59 | 56 | 53 | 50 |
| 34 | 71 | 68 | 64 | 61 | 58 | 55 | 52 | 49 |
| 35 | 70 | 66 | 63 | 60 | 56 | 53 | 50 | 47 |
| 36 | 68 | 65 | 61 | 58 | 55 | 52 | 49 | |
| 37 | 67 | 63 | 60 | 57 | 54 | 51 | | |
| 38 | 65 | 62 | 59 | 55 | 52 | | | |
| 39 | 64 | 61 | 57 | 54 | | | | |
| 40 | 62 | 59 | 56 | | | | | |
| 41 | 61 | 58 | | | | | | |
| 42 | 60 | | | | | | | |

TABLE XXXII (*c*).

| | 16 | 17 | 18 | 19 | 20 | 21 | 22 | 23 |
|---|---|---|---|---|---|---|---|---|
| 1 | 97 | 94 | 90 | 86 | 82 | 79 | 76 | 72 |
| 2 | 95 | 92 | 88 | 84 | 81 | 77 | 74 | 71 |
| 3 | 94 | 90 | 86 | 82 | 79 | 76 | 72 | 69 |
| 4 | 92 | 88 | 84 | 81 | 77 | 74 | 71 | 67 |
| 5 | 90 | 86 | 82 | 79 | 76 | 72 | 69 | 66 |
| 6 | 88 | 84 | 81 | 77 | 74 | 71 | 68 | 64 |
| 7 | 86 | 83 | 79 | 76 | 72 | 69 | 66 | 63 |
| 8 | 84 | 81 | 77 | 74 | 71 | 68 | 64 | 61 |
| 9 | 83 | 79 | 76 | 73 | 69 | 66 | 63 | 60 |
| 10 | 81 | 78 | 74 | 71 | 68 | 65 | 62 | 59 |
| 11 | 79 | 76 | 73 | 69 | 66 | 63 | 60 | 57 |
| 12 | 78 | 74 | 71 | 68 | 65 | 62 | 59 | 56 |
| 13 | 76 | 73 | 70 | 66 | 63 | 60 | 57 | 54 |
| 14 | 75 | 71 | 68 | 65 | 62 | 59 | 56 | 53 |
| 15 | 73 | 70 | 66 | 63 | 60 | 57 | 54 | 51 |
| 16 | 71 | 68 | 65 | 62 | 59 | 56 | 53 | 50 |
| 17 | 70 | 67 | 64 | 60 | 57 | 54 | 52 | 49 |
| 18 | 68 | 65 | 62 | 59 | 56 | 53 | 50 | 47 |
| 19 | 67 | 64 | 61 | 58 | 55 | 52 | 49 | 46 |
| 20 | 65 | 62 | 59 | 56 | 53 | 50 | 47 | 45 |
| 21 | 64 | 60 | 58 | 55 | 52 | 49 | 46 | 43 |
| 22 | 62 | 59 | 56 | 53 | 50 | 48 | 45 | 42 |
| 23 | 61 | 58 | 55 | 52 | 49 | 46 | 43 | 41 |
| 24 | 60 | 57 | 54 | 51 | 48 | 45 | 42 | 39 |
| 25 | 58 | 55 | 52 | 49 | 46 | 43 | 41 | 38 |
| 26 | 57 | 54 | 51 | 48 | 45 | 42 | 39 | 37 |
| 27 | 55 | 52 | 49 | 46 | 44 | 41 | 38 | 35 |
| 28 | 54 | 51 | 48 | 45 | 42 | 39 | 37 | |
| 29 | 53 | 50 | 47 | 44 | 41 | 38 | | |
| 30 | 51 | 48 | 45 | 42 | 40 | | | |
| 31 | 50 | 47 | 44 | 41 | | | | |
| 32 | 48 | 46 | 43 | | | | | |
| 33 | 47 | 44 | | | | | | |
| 34 | 46 | | | | | | | |

TABLE XXXII (*e*).

| | 32 | 33 | 34 | 35 | 36 | 37 | 38 | 39 |
|---|---|---|---|---|---|---|---|---|
| 1 | 45 | 43 | 40 | 37 | 35 | 32 | 29 | 27 |
| 2 | 44 | 41 | 39 | 36 | 33 | 31 | 28 | 25 |
| 3 | 43 | 40 | 37 | 35 | 32 | 29 | 27 | 24 |
| 4 | 41 | 39 | 36 | 33 | 31 | 28 | 25 | 23 |
| 5 | 40 | 37 | 35 | 32 | 29 | 27 | 24 | 21 |
| 6 | 39 | 36 | 33 | 31 | 28 | 25 | 23 | 20 |
| 7 | 37 | 35 | 32 | 29 | 27 | 24 | 21 | 19 |
| 8 | 36 | 33 | 31 | 28 | 25 | 23 | 20 | 18 |
| 9 | 35 | 32 | 29 | 27 | 24 | 21 | 19 | 16 |
| 10 | 33 | 31 | 28 | 25 | 23 | 20 | 18 | 15 |
| 11 | 32 | 29 | 27 | 24 | 22 | 19 | 16 | 14 |
| 12 | 31 | 28 | 25 | 23 | 20 | 18 | 15 | |
| 13 | 29 | 27 | 24 | 22 | 19 | 16 | | |
| 14 | 28 | 25 | 23 | 20 | 18 | | | |
| 15 | 27 | 24 | 22 | 19 | | | | |
| 16 | 26 | 23 | 20 | | | | | |
| 17 | 24 | 22 | | | | | | |
| 18 | 23 | | | | | | | |

TABLE XXXII (*d*).

| | 24 | 25 | 26 | 27 | 28 | 29 | 30 | 31 |
|---|---|---|---|---|---|---|---|---|
| 1 | 69 | 66 | 63 | 60 | 57 | 54 | 51 | 48 |
| 2 | 67 | 64 | 61 | 58 | 55 | 52 | 50 | 47 |
| 3 | 66 | 63 | 60 | 57 | 54 | 51 | 48 | 45 |
| 4 | 64 | 61 | 58 | 55 | 52 | 50 | 47 | 44 |
| 5 | 63 | 60 | 57 | 54 | 51 | 48 | 45 | 43 |
| 6 | 61 | 58 | 55 | 53 | 50 | 47 | 44 | 41 |
| 7 | 60 | 57 | 54 | 51 | 48 | 45 | 43 | 40 |
| 8 | 58 | 55 | 52 | 50 | 47 | 44 | 41 | 39 |
| 9 | 57 | 54 | 51 | 48 | 46 | 43 | 40 | 37 |
| 10 | 56 | 53 | 50 | 47 | 44 | 41 | 39 | 36 |
| 11 | 54 | 51 | 48 | 46 | 43 | 40 | 37 | 35 |
| 12 | 53 | 50 | 47 | 44 | 41 | 39 | 36 | 33 |
| 13 | 51 | 48 | 46 | 43 | 40 | 37 | 35 | 32 |
| 14 | 50 | 47 | 44 | 42 | 39 | 36 | 33 | 31 |
| 15 | 49 | 46 | 43 | 40 | 37 | 35 | 32 | 29 |
| 16 | 47 | 44 | 42 | 39 | 36 | 33 | 31 | 28 |
| 17 | 46 | 43 | 40 | 37 | 35 | 32 | 29 | 27 |
| 18 | 44 | 42 | 39 | 36 | 33 | 31 | 28 | 26 |
| 19 | 43 | 40 | 38 | 35 | 32 | 30 | 27 | 24 |
| 20 | 42 | 39 | 36 | 34 | 31 | 28 | 26 | |
| 21 | 40 | 38 | 35 | 32 | 30 | 27 | | |
| 22 | 39 | 36 | 34 | 31 | 28 | | | |
| 23 | 38 | 35 | 32 | 30 | | | | |
| 24 | 36 | 34 | 31 | | | | | |
| 25 | 35 | 32 | | | | | | |
| 26 | 34 | | | | | | | |

TABLE XXXII (*f*)

| | 40 | 41 | 42 | 43 | 44 | 45 | 46 | 47 | 48 | 49 |
|---|---|---|---|---|---|---|---|---|---|---|
| 1 | 24 | 21 | 19 | 16 | 14 | 11 | 09 | 06 | 04 | 01 |
| 2 | 23 | 20 | 18 | 15 | 13 | 10 | 08 | 05 | 03 | |
| 3 | 21 | 19 | 16 | 14 | 11 | 09 | 06 | 05 | | |
| 4 | 20 | 18 | 15 | 13 | 10 | 08 | 05 | | | |
| 5 | 19 | 16 | 14 | 11 | 09 | 06 | | | | |
| 6 | 18 | 15 | 13 | 10 | 08 | | | | | |
| 7 | 16 | 14 | 11 | 09 | | | | | | |
| 8 | 15 | 13 | 10 | | | | | | | |
| 9 | 14 | 11 | | | | | | | | |
| 10 | 13 | | | | | | | | | |

With the aid of Table XXXII. one can turn measurements by relative position into measurements in units of + or — $\sigma$ almost as fast as one can read.

For instance, of 800 schoolboys,

| | | | |
|---|---|---|---|
| 8 per cent. | received a mark of | | *E* |
| 20 per cent. | " | " | *VG* |
| 38 per cent. | " | " | *G* |
| 24 per cent. | " | " | *F* |
| 8 per cent. | " | " | *P* |
| 2 per cent. | " | " | *U* |

The table tells us at once that, in so far as the distribution of the ability in the group in question follows the type of distribution described above,

$$
\begin{aligned}
E &= +1.86\,\sigma \\
VG &= +\ .93\,\sigma \\
G &= +\ .08\,\sigma \\
F &= -\ .79\,\sigma \\
P &= -1.58\,\sigma \\
U &= -2.44\,\sigma
\end{aligned}
$$

There is still another possibility of turning measures by relative position into units of amount and so making them available for common scientific usage. In certain cases it may be justifiable to suppose that the least noticeable difference is a constant quantity for any one trait for any one observer; in simpler words, that if I say that John, James and Peter are to me indistinguishable, say, in literary merit, but that Henry and William are a shade better, and that George and Fred are a shade better than Henry and William, the actual difference between *JJP* and *HW* equals that between *HW* and *GF*. In so far as this were true we could, with a large group of individuals varying continuously from the low to the high extreme, make groups on the basis of the least noticeable difference and call the steps of ability from group to group always the same.

The measures are then identical in form with those by ordinary units of amount. The only difference is that the amount of the quantity at the starting-point of the whole group ($A$) and the amount of the step from one subgroup to the next ($K$) are unknown except from the things measured themselves and are undefinable except in terms of them. The question, 'How much are $A$ and $K$?' can be answered only by pointing to the achievements of the lowest group and saying, 'That is $A$,' by pointing to the differences between that group and others and saying, 'This much difference is $K$, this much $4K$, this much $20K$ and so on.'

The hypothesis that the least noticeable difference in a trait is for the same observer a constant quantity has not been tested sufficiently to decide how far its use is justifiable, but there can be no doubt that some modification of the principle involved will sometime be a valuable resource of the theory of mental measurements.

For the sake of simplicity, only the case of individuals measured by their relative position in a group has been discussed in this chapter. Everything in the chapter applies equally well to measures of the different trials of one individual.

## PROBLEMS.

21. Turn into statements in units of the A. D. of the distribution, measured + and − from the average, the measures by relative position given below; first, on the supposition that the form of distribution is a rectangle; second, on the supposition that the form of distribution is of the normal type (use Table XXXII.); third, with no supposition about the form of distribution, but on the hypothesis that the measures represent a grouping by the least noticeable differences and that these differences are equal:

*A*, *B*, *C*, *D*, *E* and *F* are marks running from high to low. Of some 200 and over high-school students, 2 per cent. received *A*, 22 per cent. *B*, 44 per cent. *C*, 25 per cent. *D*, 6 per cent. *E*, and 1 per cent. *F*.

22. Which supposition is the more likely? Why?

23. Using Table XXXI., calculate the measure in terms of units of amount (1) of the highest four per cent. of a group normally distributed; (2) of the six per cent. just above the mode; (3) of the three per cent. from the end of the seventeenth down, *i. e.*, of per cents. 18th, 19th and 20th. Verify the results from the entries for these groups in Table XXXII.

24. On the hypothesis that the distribution of darkness of eyes is normal, use Table XXXII., and transmute into terms of units of amount the following measures by relative position:

| Eye Color. | Per Cents. of Englishmen.* | | |
|---|---|---|---|
| Light blue. | 2.9 | call | 3 |
| Blue. Dark blue. | 29.3 | " | 29 |
| Gray. Blue-green. | 30.2 | " | 30 |
| Dark gray. Hazel. | 12.3 | " | 12 |
| Light brown. Brown. | 11.0 | " | 11 |
| Dark brown. | 10.8 | " | 11 |
| Very dark brown. Black. | 3.6 | " | 4 |

It is possible to use the table for a finer scale than to a single per cent. by interpolating. But it is hardly worth while.

* From Galton's 'Natural Inheritance.'

# CHAPTER VIII.

## THE MEASUREMENT OF DIFFERENCES AND OF CHANGES.

THE chief questions that concern the measurement of differences in the mental sciences arise in the case of comparisons of groups and measurements of changes. Instead of any general abstract treatment of the measurement of differences, therefore, I shall present the special applications of it to these two problems. Only a very brief outline of the problem as a whole will be given as an introduction.

The difference between any two amounts of the same kind of fact may be measured. The amounts may be:

1. Two single figures, each standing for a general tendency, *e. g.*, averages, medians or modes.

2. Two single figures, each standing for a variability, *e. g.*, A. D.'s, $\sigma$'s or P. E.'s.

3. Two single figures, each standing for a difference itself.

4. Two single figures, each standing for a relationship.

5. Two total distributions, each standing for a general tendency plus the deviations from it.

The general tendency may be to the possession of a certain amount of variability, of difference or of relationship, as well as of a thing or quality. It will, however, commonly be the latter.

The classification above could, of course, be extended *ad infinitum* with such complexities as: "The measurement of the difference between two variabilities, each being of the amounts of relationship between the amount of difference between (1) 10-year-olds and 11-year-olds in motor ability and (2) 10- and 11-year-olds in sense discrimination."

The difference between two single figures will be measured (*a*) by the gross difference; (*b*) by the per cent. the difference is of the amount of one of them.

The difference between two total distributions will be measured fully by comparing them item by item; the measurement may be summarized in various ways.

The difference between two facts, each of which is measured by its relative position in a series, may be measured most satisfactorily by transmuting the series and then using regular methods, most quickly by the gross or percentile difference between the two, rated as members of the same series.

*The Comparison of Groups.*

The common custom of comparing groups by comparing their averages is inadequate because for both practical and theoretical purposes the meaning of a difference between two averages depends upon the variabilities of the groups. The mere fact, for example, that in the *A* test (see page 46) the averages for 12-year-old boys and for 12-year-old girls were respectively 41 and 46, might mean (1) that the lowest ranking girl was above the highest ranking boy, *i. e.*, that boys and girls were in this trait totally distinct species or (2) that only 5 per cent. of girls were better than the highest ranking boy, or even (3) that no girl was equal to the highest ranking boy. It might mean, in fact, all sorts of conditions, some of which are pictured in Figs. 71 to 76.

It is of no great advantage to estimate the difference in a per cent. rather than a gross amount. One group may in ten different tests have always an average twenty per cent. higher than the other, and yet the differences in ability may really be equal in no two of the ten cases. For, since in mental and social traits there are rarely absolute zero points at which to start the scale,* the meaning of each percentage will depend upon the number chosen as the starting-point in measuring. We can always make a difference so expressed seem less by starting the scale at 10 or 40 or 100 instead of at 0 or 4 or 10. And the same percentage in a case where the variability of the trait is great will always mean for practical purposes a less difference than it does in a case where the variability is small.

For instance, if the *A* test is scored by the number of *A*'s marked, the percentage superiority of girls to boys is 12.2 ; if by the number marked more than the lowest 12-year-old record, it is 18.5 ; if by the number of *A*'s omitted, it is 8.5. Clearly the figure depends on an entirely arbitrary factor.

What is needed for the comparison of groups is some measure

* See Chapter II., pp. 15 and 16.

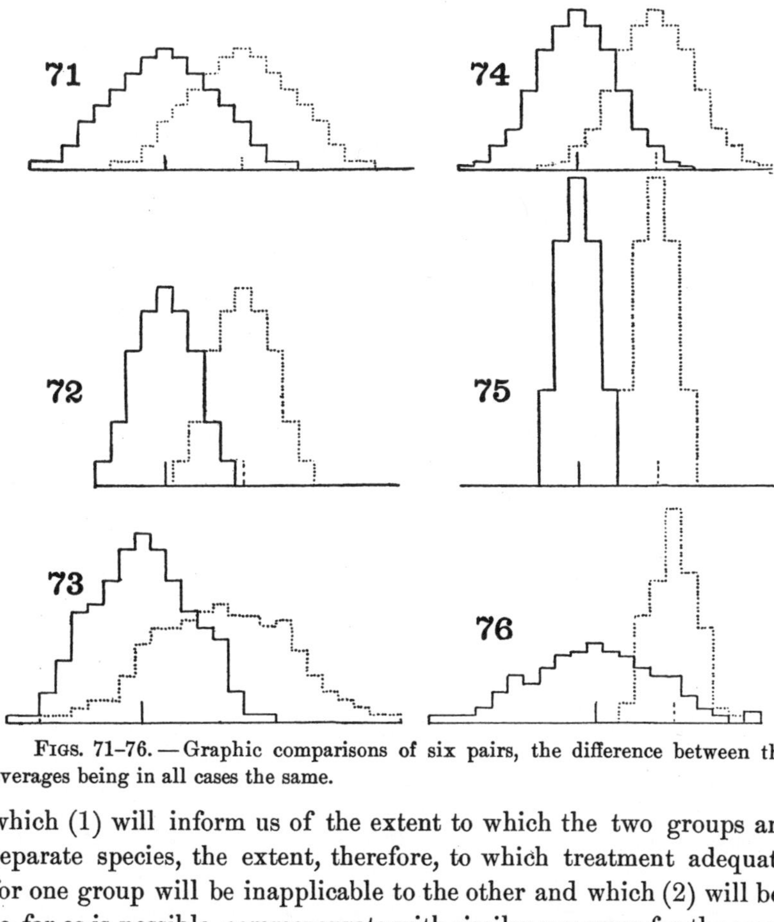


FIGS. 71–76. — Graphic comparisons of six pairs, the difference between the averages being in all cases the same.

which (1) will inform us of the extent to which the two groups are separate species, the extent, therefore, to which treatment adequate for one group will be inapplicable to the other and which (2) will be, so far as is possible, commensurate with similar measures for the same groups in other traits, so that we may compare the differences of groups in different traits.

The first desideratum is met by comparing the two total distributions instead of the mere averages, or approximately in the case of traits somewhat normally distributed, by stating the variabilities of the two groups. Thus, to use our previous illustration, the distribution of 12-year-old boys and of 12-year-old girls in the *A* test as given in Table XXXIII. and Fig. 77, tells us at once that the difference between the averages is 5.2, that over 99 per cent. of the girls are contained between the same limits of ability as the boys, that only 31 per cent. of boys reach the median mark for girls, that the sex difference is far less important practically than individual differences within either sex, that between 28 and 62 are 88.7 per cent of the boys and 87.4 per cent. of the girls. These same measures could be obtained approximately from the theoretical properties of the normal surface of frequency if the variabilities of the groups were given instead of the total distributions.

The second desideratum is met by measuring the difference in terms of the per cent. of one group who reach or exceed the median mark for the other group (or some other set measure). If in Latin,

TABLE XXXIII.

*A*'s Marked in 60 Seconds.

| Quantity. | Frequency. | |
|---|---|---|
| | 12-year-old boys. | 12-year-old girls. |
| 14 — 16 | | 1 |
| 16 — 18 | 2 | |
| 18 | 1 | 1 |
| 20 | | |
| 2 | 4 | 2 |
| 4 | 4 | 1 |
| 6 | 3 | 2 |
| 8 | 9 | 1 |
| 30 | 10 | 2 |
| 2 | 8 | 4 |
| 4 | 10 | 11 |
| 6 | 15 | 5 |
| 8 | 15 | 9 |
| 40 | 10 | 11 |
| 2 | 13 | 9 |
| 4 | 12 | 14 |
| 6 | 13 | 10 |
| 8 | 8 | 7 |
| 50 | 4 | 6 |
| 2 | 6 | 7 |
| 4 | 3 | 6 |
| 6 | 2 | 4 |
| 8 | 1 | 8 |
| 60 | 4 | 4 |
| 2 | 1 | 4 |
| 4 | 1 | 3 |
| 6 | | 1 |
| 8 | | 1 |
| 70 | | |
| 2 | | 1 |
| 4 | 1 | |
| 6 | | |
| 8 | 1 | |

Greek, algebra and history one group of students always show 30 per cent., reaching the median of another group, then it is true to say that the second group is equally superior in all four of these studies. At least there can be no better evidence of equality in amount of difference in mental traits than this.

Under the present conditions of thoughtless measurements of mental traits it frequently happens that groups will be compared with respect to the same trait by different tests, and no one will be able

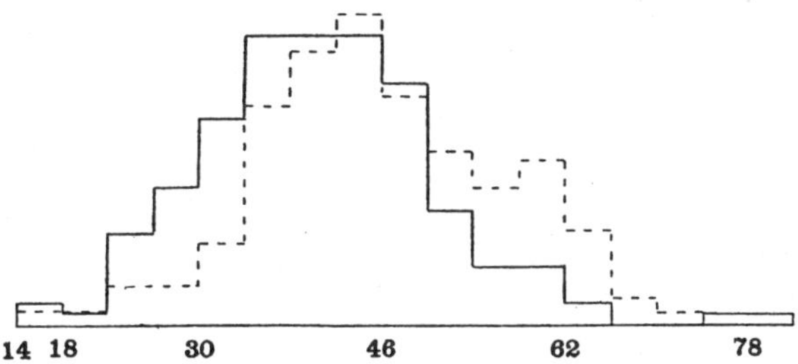


FIG. 77. — The continuous line gives the distribution of ability in perception (*A* test) in 12-year-old boys; the dotted line that for girls. The cases are grouped more coarsely than in the table.

to tell how far results agree. If the mere averages were replaced by the measure *per cent. of group 1 reaching median of group 2*, results by all sorts of methods could be put together. It is, of course, true that when one group so far exceeds another that its lowest score is above the highest score of the other, the method suggested here fails. Such cases are, however, extremely rare in the comparisons of groups characterized by differences of sex, training, age, social conditions, birth, occupation, locality, etc., such as psychology, education and sociology are studying.

In these cases of total disparity in the two distributions, the results from different tests may be made commensurate, so far as is possible, by expressing the differences in terms of the variability of one of the two groups.

Comparison by the per cent. of one group that exceed the median measure of some other group has the further advantage of being applicable to groups measured by relative position only. For instance, if one knew that the crimes in one town were as listed in column 1, and those of a second town as listed in column 2, he could state that almost 59 per cent. of the first town's crimes were greater than the median crime of the second, could thus have a quantitative comparison of the two without having to adopt speculative equivalents of one crime in terms of others.

| Offense. | Frequency in first town. | Frequency in second town. |
|---|---|---|
| Peddling without a license | 2 | 3 |
| Failure in jury duty | 4 | 5 |
| Disturbing the peace | 9 | 11 |
| Drunkenness | 23 | 28 |
| Robbery | 30 | 27 |
| Assault and robbery | 17 | 11 |
| Arson | 8 | 10 |
| Murder in second degree | 5 | 4 |
| Murder in first degree | 1 | 1 |
| Patricide | 1 | |

In comparing groups with respect to variability, allowance must be made for the fact that the amount of the central tendency influences the size of the $\sigma$ or A. D. or P. E. that is obtained. For instance, 22 individuals added for 40 seconds, and gave a group score of—Median, 9.0; A. D., 2.18. The same 22 individuals then added for 80 seconds and gave a group score of—Median, 16.0; A. D. 3.41. In a final test for 120 seconds, the results were—Median, 23.5; A. D., 5.18. These figures do not mean that the real variability of the group doubled within a few minutes, or that it altered at all, but only that the gross amount of the variability depends upon the gross amount of the measures themselves as well as upon the real variability. The gross amount of variability in the length of the line drawn by a group of individuals trying to equal a 10-mm. line will be far less than the gross variation of their attempts to equal a 1,000-mm. line, yet the real variability is presumably the same.

Just how much allowance to make it is difficult to decide. Karl Pearson has proposed, as a measure of variability by which groups may be fairly compared, the gross variability divided by the average. By this figure, which we may call the Pearson Coefficient of Variability, we should, in the case of the 12-year-old boys and girls in the *A* test (Boys, Av. 40.7, A. D., 8.1; Girls, Av. 45.9, A. D., 8.5) reverse the gross difference, the girls becoming only 93 per cent. as variable as the boys. It would seem to the author more in accord with both theory and facts to use the gross variability divided by the square root of the average. Any such comparison is misleading if there are no real, but only arbitrary, zero points.

Comparisons of groups in variability are of two sorts: (1) Of different groups with respect to their variabilities in the same trait. (2) Of the same group with respect to its variabilities in different traits.

In the first case the differences between the averages in the cases which interest the student are commonly not very great, and the zero points, though arbitrary, are subject to not very great fluctuations; consequently the comparison by any method is commonly such as to reveal any marked difference in variability that exists. In practice one can do no more than present the two total distributions the variabilities of which are to be compared, explain what zero points were taken and why, and calculate for the reader the relation of the group's variabilities by all three methods. Often it is best simply to present the gross variability and leave any one to allow for differences in the amount of the measures themselves as he sees fit.

The second case will only rarely be an important object of study. This is fortunate, since here the differences between averages may run to any amount, and the zero points for some of the traits may be subject to extreme variations. For instance, suppose that one wished to compare the variabilities of adult men in salary, morality, health and intellect. The average of the first may be 600; that of the second, 10; that of the third, 1,000, and that of the fourth, 10,000, according to the units and zero points chosen. We would take as our zero point for salary $0.00 per year, but some men are actually a burden and should be rated as minus. The absolute zero point, then, some one may put at the point of the man whose work is worth nothing to any one and whose care costs the most. So also morality may be reckoned upward from the lowest clergyman or from the lowest criminal. Again, is the zero point for health that of one who just keeps above dying for a moment, or that of the sickest one found in the group?

In practice one can do no more than to present the total distributions, explain what zero points were taken and why, and use proper logic in inferring anything about the relations of the variabilities found.

*The Measurement of Changes.*

By a change in anything is meant the difference between two conditions of it. It might seem that the problem of the measurement of changes was identical with that of measuring differences, and that this section was superfluous. In a certain sense this is true.

The general principles of previous chapters do answer the special questions of this chapter. But it will be clearer, and in the end save the student's time, to study these special questions separately, especially since in studies of change one is commonly concerned with a number of successive steps of difference, and is trying to measure, not a single alteration, but a continuous process of alteration.

*The Measurement of a Change in an Individual.*

A mere series of averages does not give the data for a complete measurement of the change. The averages might be the same and yet the constancy of performance of the individual might have altered. Thus the average values of a stock from 1890 to 1900 might be alike and yet it might have changed from a fluctuating uncertainty in 1890, with say, an average deviation of 40, to a steady assured value in 1900, with an average deviation of only 3. The stock in 1890 would be more desirable property than the stock in 1900 from the point of view of one moved by the gambler's instinct; the reverse would hold for a steady-going man with a family or for a conservative bank. To measure change fully one needs a series of total distributions. If they are not at hand one must be sure not to pretend to measure something other than that represented by the series of quantities he does have.

Inequalities in units are more likely to escape attention in measurements of change than anywhere else. Yet it is just in such measurements that they may do the most harm. For instance, all statistics with which I am acquainted measure the change in the death-rates from various diseases by series of figures, each giving the proportion of deaths to cases or to total population or to some other standard, as in the following: *

In 1891, 22.5 per cent. of those having diphtheria died.
" 1892, 22.2 " " " " " "
" 1893, 23.3 " " " " " "
" 1894, 23.6 " " " " " "
" 1895, 20.4 " " " " " "
" 1896, 19.3 " " " " " "
" 1897, 17.0 " " " " " "
" 1898, 14.8 " " " " " "
" 1899, 14.2 " " " " " "
" 1900, 12.8 " " " " " "

* 'London Statistics,' Vol. XII., p. 97 of the Medical Officer's Report.

But such figures can not be taken at their face value, for to cure one case of diphtheria is not the same quantity of progress as to cure another. The progress of medicine and hygiene which reduces the death-rate from 40 to 30 does so presumably often by curing the easiest quarter of those previously uncured. The next cases will be harder, and possibly to cure the last 1 per cent. of the 40 would mean more advance in medicine and hygiene than was needed for the curing of all the other 99.

When the change is in number of individuals affected or number of errors made or number of tasks done, there is then special danger

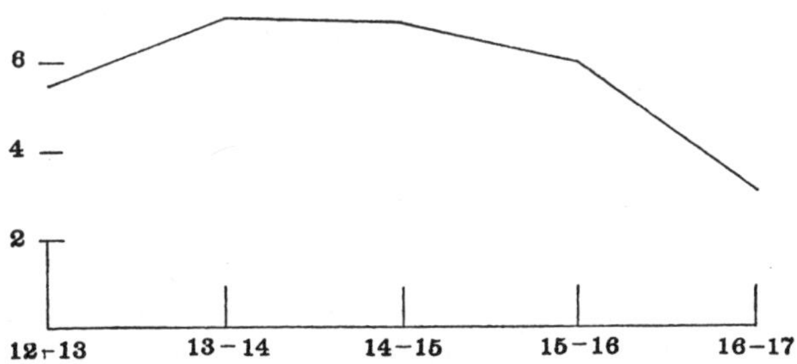


FIG. 78. — The heights of the line above the base line at the points 12–13. 13–14, 14–15, 15–16, 16–17, give the differences between the average height at 12 and that at 13, the difference between the average height at 13 and that at 14, etc., for 25 boys measured annually for five years.

in neglecting the inequalities among the units; for the change will commonly single out the easiest first.

The common absence of zero points in the case of mental measurements makes it unwise to express changes in percentile increments, and definitely unjustifiable to so express them if the gross amounts whence the percentages are derived are not also given. If, for instance, I am informed that *A*'s reaction time improved 10 per cent. per year from 6 to 12 years, I am at a loss to tell what is meant.

In comparing two (or more) individuals with respect to change one may use gross change, percentile change or change in terms of the variabilities of the individuals, provided that he makes it clear which he is using and, of course, treats both individuals alike. No one method is the correct one; all are correct, but measure different things. 4 to 5 equals 8 to 9 if by change is meant amount added; 4 to 5 equals 8 to 10 if one means proportion added; 4 to 5 (the A. D. of 4 being 2) equals 8 to 9.5 (the A. D. of 8 being 3) if one means distance traversed toward the extreme ability of the previous condition. This is all that can be said in general. Each special case may offer reasons for preferring one method. The beginner in statistical work may well use all three.

### *The Measurement of a Change in a Group.*

This heading is ambiguous in that it may be taken to refer: (1) to the measurement of the changes undergone by a series of individuals, or (2) to the change undergone by some measure of a group. It should be needless to say that the two questions are radically different, but they are often confused. The changes in stature of 100 boys from the age 15 to the age 16 are not the change from the average stature of the group 100 boys at 15 to the average stature of the same group at 16 years. The first fact, the total fact of all the individal changes, is calculated from 100 individual measures of change, is a distribution with an ascertainable variability and in all respects stands in the same relation to individual changes as does the distribution of an ability in a group to the abilities of its members. The second fact is calculated as the difference of two averages, has no known variability, is, in fact, simply a partial measure of difference between two groups. If our argument is ever to return to individual changes, the first sort of measure must be used. This will commonly be the case.

For an example take the case of the change in stature of 25 boys from the twelfth to the seventeenth year.* If we try to infer anything about growth from the change in average stature, we have only the following facts: Average stature for 12, 13, 14, 15, 16 and 17 year old boys, 142.6, 148.12, 154.92, 161.60, 167.64 and 170.76 centimeters respectively. Yearly differences, + 5.52, + 6.8, + 6.68, + 6.04 and + 3.12 centimeters. These differences are shown in Fig. 78.

If, on the other hand, we preserve the individual changes in our statement, we have the facts of Table XXXIV.

These show the great variability in growth and the law of compensation that 'boys who were tall at 12 years grow the faster during the interval 12 to 13 and 13 to 14; but during the intervals of 14

*For these measurements I am indebted to the kindness of Professor Franz Boas and Dr. Clark Wissler.

to 15 and 15 to 16 they grow slowly; with the boys of short stature at 12 the rates of growth are exactly the reverse.'* How the single yearly differences above fail to represent the real complexity and correlation of the facts can be seen by comparing Fig. 78 with Fig. 79, which shows the real changes of the 25 individuals. Fig. 80 brings out more clearly the inverse relation between the change from 12 to 14 and that from 14 to 16.

TABLE XXXIV.

GROWTH OF 25 BOYS FROM THE 12TH THROUGH THE 17TH YEAR.

| | Change. | | | | |
|---|---|---|---|---|---|
| Stature at 12. | 12–13. | 13–14. | 14–15. | 15–16. | 16–17. |
| 132 | 5 | 7 | 10 | 6 | 4 |
| 134 | 5 | 5 | 7 | 10 | 3 |
| 135 | 5 | 2 | 6 | 8 | 8 |
| 135 | 6 | 7 | 10 | 6 | 4 |
| 136 | 4 | 8 | 8 | 7 | 2 |
| 136 | 7 | 9 | 5 | 3 | 2 |
| 137 | 4 | 6 | 8 | 6 | 4 |
| 137 | 5 | 4 | 8 | 10 | 5 |
| 139 | 4 | 8 | 7 | 7 | 2 |
| 140 | 5 | 7 | 10 | 6 | 3 |
| 142 | 9 | 7 | 6 | 3 | 1 |
| 142 | 4 | 5 | 5 | 10 | 6 |
| 143 | 4 | 5 | 5 | 8 | 7 |
| 144 | 6 | 11 | 6 | 3 | 1 |
| 145 | 6 | 5 | 7 | 8 | 4 |
| 146 | 4 | 4 | 6 | 10 | 4 |
| 146 | 4 | 7 | 8 | 3 | 1 |
| 146 | 4 | 5 | 4 | 11 | 2 |
| 146 | 9 | 11 | 5 | 2 | 1 |
| 147 | 4 | 7 | 9 | 5 | 4 |
| 147 | 8 | 10 | 7 | 3 | 1 |
| 149 | 7 | 13 | 1 | 5 | 1 |
| 151 | 5 | 7 | 8 | 3 | 4 |
| 152 | 5 | 4 | 10 | 5 | 2 |
| 158 | 9 | 6 | 1 | 3 | 2 |

For the measurement of change in a group (that is, of all the individual changes), the statistical treatment is, as suggested above, simply that for any fact in a group, the fact here being an amount of change instead of an amount of a thing or condition. The need of a statement in a table of frequencies and the use of average, mode, median and the various measures of variability — in fact, the entire theory of Chapters III. and IV. — are applicable here.

* 'The Growth of Boys,' by Clark Wissler, *American Anthropologist* (New Series), Vol. 5, pp. 83 and 84.

For the measurement of change from one condition of a group to another the statistical treatment is simply that described in the case of the measurement of difference.

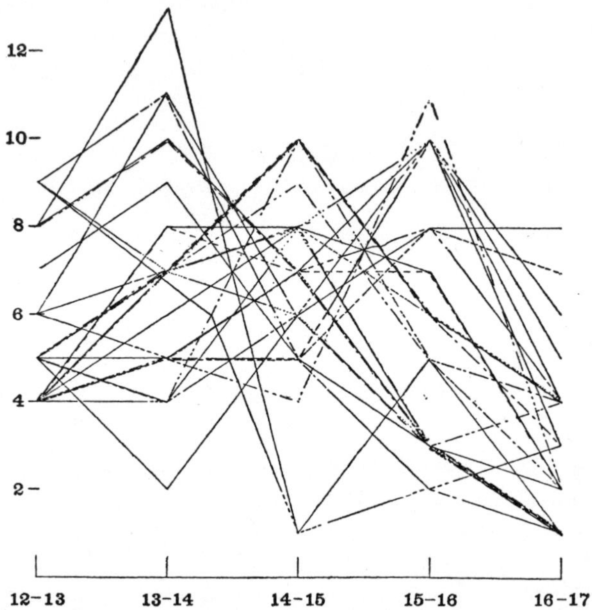


FIG. 79. — The heights of the five points *A*, *B*, *C*, *D*, *E* of each line measure the yearly differences for one individual as did the line of Fig. 78 the yearly differences for the average stature of the group. The figure, that is, presents graphically the facts of Table XXXIV.

## PROBLEMS.

25. In which trait, *A* or *B*, is there the greater difference between Group I. and Group II ?

| Quantity *A*. | Frequency. Group I. | Frequency. Group II. | Quantity *B*. | Frequency. Group I. | Frequency. Group II. |
|---|---|---|---|---|---|
| 39 | 1 | 1 | 2 | 1 | 5 |
| 40 | 1 | 0 | 3 | 0 | 5 |
| 41 | 4 | 1 | 4 | 3 | 6 |
| 42 | 11 | 7 | 5 | 7 | 6 |
| 43 | 23 | 16 | 6 | 13 | 8 |
| 44 | 25 | 20 | 7 | 20 | 10 |
| 45 | 28 | 22 | 8 | 22 | 16 |
| 46 | 28 | 26 | 9 | 15 | 10 |
| 47 | 30 | 30 | 10 | 3 | 8 |
| 48 | 20 | 27 | 11 | 3 | 8 |
| 49 | 9 | 18 | 12 | 1 | 6 |
| 50 | 3 | 10 | 13 | 1 | 7 |
| 51 | 0 | 5 | 14 | | 2 |
| 52 | 1 | 2 | 15 | | 3 |

26. Groups III. and IV. are approximately normally distributed. Group III. has Median = 10 and A. D. = 4 and Group IV. has Median = 12 and A. D. = 3. What per cent. of Group IV. will exceed the median for Group III.? What per cent. of Group III. will exceed the median for Group IV.? (The table on page 60 affords the further data necessary.)

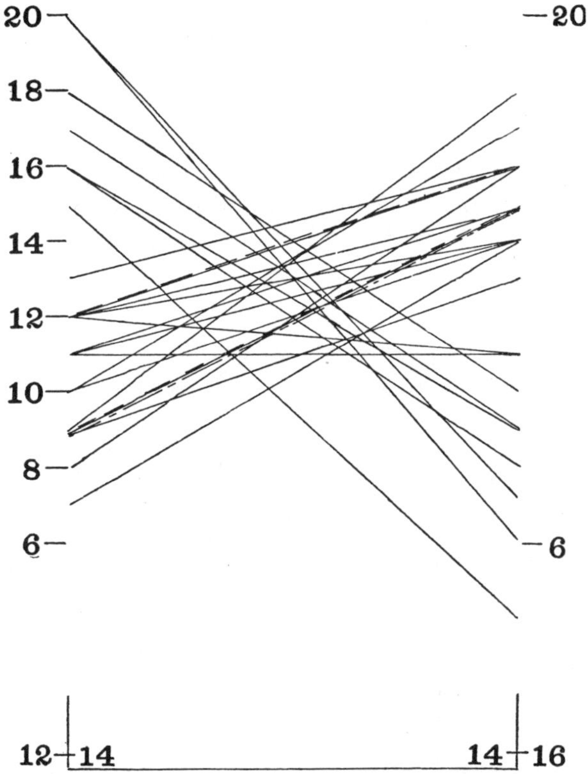


FIG. 80. — The height of any one of the lines at its left-hand extreme measures the change in stature of one boy from 12 to 14; its height at the right hand extreme measures the change from 14 to 16.

27. If we know the average wealth of 100 men in 1900 to be $5,000 and in 1905 to be $10,000, what do we know about the changes that have taken place?

28. Recall any arguments based on the application to individuals of some change true of them only as a group. Where else have we in this book met a similar fallacy?

## CHAPTER IX.

### THE MEASUREMENT OF RELATIONSHIPS.

THE difficulty of measuring mental and social relationships is, of course, due to their variability. The relation of the weight of a gas at constant temperature and pressure to its volume we assume to be always the same, but the relation of intellect to morality is almost never the same; the relation of the force of gravity to the product of the masses of the two bodies is constant, but the relation of ability in school to efficiency in life is very variable. The problem is thus to represent the total tendency shown by many different individual relationships.

*Case I.*

The relationship of changes in the amount of one thing to changes in the amount of another thing, when the things are physical, is shown by a series of corresponding values of the two things reckoned from zero points in both cases, each pair of values being represented by two constants. It is expressed mathematically by the equation which represents the way in which the amount of the one thing depends upon the amount of the other.

The following case may serve as an illustration:

$n =$ the index of refraction of air.

$d =$ the density of air.

$p$ (a quantity subject to the control of the experimenter) $= C_1 d$.

$N$ (a quantity measurable by the experimenter) $= C_2(n - 1)$.

$C_1$ and $C_2$ are constants.

The experiments consisted in varying $p$ and measuring the related changes in $N$. The results are as follows:

| | | | | | | |
|---|---|---|---|---|---|---|
| When | $p$ | is | 9.989 | $N$ | is | 316.7 |
| " | " | " | 10.146 | " | " | 321.2 |
| " | " | " | 10.163 | " | " | 321.6 |
| " | " | " | 18.281 | " | " | 579.2 |
| " | " | " | 18.365 | " | " | 582.7 |
| " | " | " | 26.932 | " | " | 852.6 |
| " | " | " | 35.990 | " | " | 1142.1 |
| " | " | " | 48.780 | " | " | 1545.1 |

If each of these pairs of related values is turned into an equation of the form $N = xp$, the results are:

| | |
|---|---|
| $N = 31.70p$ | $N = 31.72p$ |
| $N = 31.66p$ | $N = 31.66p$ |
| $N = 31.64p$ | $N = 31.69p$ |
| $N = 31.68p$ | $N = 31.68p$ |

Obviously, a single equation $N = 31.68p$ expresses very closely the relationships found for different values of $p$.

The measurements of relationship here are, of course, not absolutely free from variability. For instance, the 10.163 came really from 7 measurements with an average deviation of .012. But the

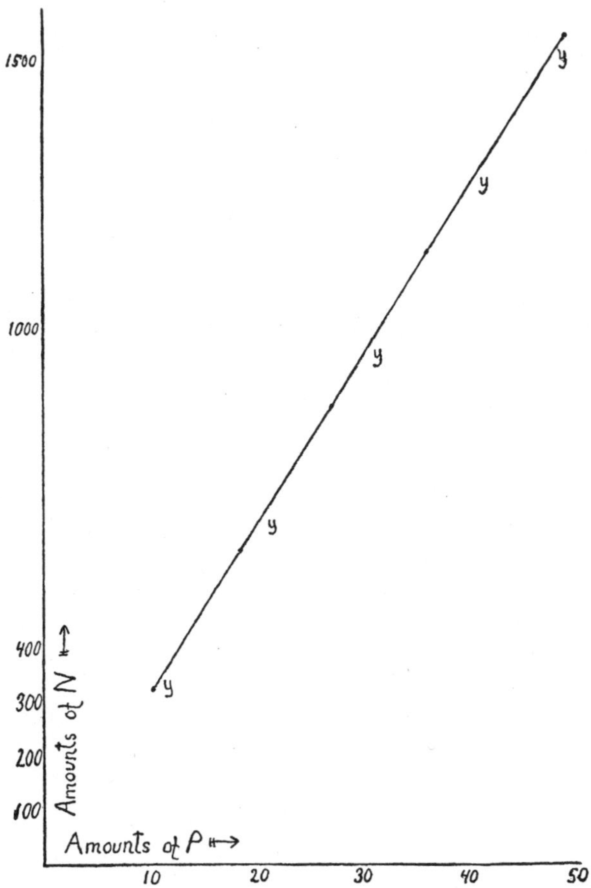


FIG. 81.

variability is here small and presumably due entirely to variations in the instruments or observers.

If the pairs of values are plotted as in Fig. 81, the slope of the line shows the relationship. The equation $N = 31.68p$ expresses very closely the slope of this line referred to its coordinates. $N/p$ is thus constant. $(n - 1)/d$ equals $N/p$ times some constant. Therefore, $(n - 1)/d$ itself equals a constant. The relation between the index of refraction of air and its density is then such that $(n - 1)/d = k$ or $n = kd + 1$.*

*Case II.*

When changes in the amount of a mental trait are to be related to changes in the amount of a physical trait, the series will be of pairs, of which one will be a constant and a quantity measured from a zero point and the other a variable and often a quantity with no ascertained zero point. The following case may serve as an illustration :

Ebbinghaus in studying the relation between the lapse of time and memory found that if a series of syllables was memorized and then 24 hours allowed to pass, there was required to rememorize the series 73.6 per cent. as much time as was originally needed. In another test, however, the result was 60.4 per cent., and he quite properly announces not only the average of all the numerous varying results, but also each separate one. So also for the time taken after intervals of 19, 63 and 525 minutes and 2 and 6 days. In the statement of the relationship which follows (in Table XXXV.), the 'time saved in learning' quite evidently is a variable. One may note the wisdom of the investigator in measuring the change, not in the ambiguous units of so many words lost, but in 'per cent. of original time taken to relearn,' a system of units with an intelligible zero point.

If we plot the pairs of values as in the previous illustrations, the result is Fig. 82, which shows the general tendency of the relationship and at the same time its lack of uniformity.

In such cases it is common to replace the tables of frequencies for the mental trait by their averages. This procedure never fully

* The figures in this illustration are quoted from a report by Henry G. Gale of a research 'On the Relation between Density and Index of Refraction of Air.' *Physical Review*, January, 1902.

describes the relationship and, unless the distributions are symmetrical about a central mode, may misrepresent it. At all events, the total fact of the relationship should always be presented, as well as its abbreviated and more convenient form. In so far as the zero point from which the mental trait is measured is unknown, it is necessary to replace all face values $y$, $y_1$, $y_2$ etc., of the mental traits measured by $k+y$, $k+y_1$, $k+y_2$, etc. The formulation of any algebraic expression for the relationship is thus less simple.

TABLE XXXV.

RELATION BETWEEN LAPSE OF TIME AND MEMORY.*

| 0.32 hrs. | 1.05 hrs. | 8.75 hrs. | 24 hrs. | 48 hrs. | 144 hrs. | 744 hrs. | |
|---|---|---|---|---|---|---|---|
| 64.3 | 49.6 | 36.0 | 26.4 | 17.4 | 21.0 | 26.0 | 20.0 |
| 55.9 | 37.4 | 29.0 | 39.6 | 32.7 | 31.1 | 31.6 | 19.4 |
| 56.6 | 47.4 | 28.0 | 35.4 | 12.3 | 32.7 | 34.7 | 22.9 |
| 62.5 | 46.8 | 30.4 | 39.9 | 28.9 | 24.4 | 31.6 | 6.7 |
| 60.7 | 51.4 | 39.8 | 34.9 | 30.6 | 17.7 | 30.3 | 6.9 |
| 63.1 | 49.1 | 35.6 | 38.9 | 46.0 | 5.9 | 20.5 | 25.9 |
| 59.1 | 44.5 | 48.2 | 46.7 | 23.5 | 34.1 | 10.1 | 18.9 |
| 56.0 | 54.5 | 31.6 | 16.7 | 25.4 | 33.3 | 6.8 | 20.5 |
| 64.4 | 42.3 | 35.5 | 21.3 | 18.4 | 28.7 | 6.5 | 11.4 |
| 44.7 | 40.9 | 40.1 | 38.6 | 23.4 | 23.2 | 13.3 | 17.3 |
| 53.6 | 34.2 | 37.9 | 29.0 | 41.0 | 40.3 | 17.7 | 17.1 |
| 57.7 | 45.4 | 38.0 | 37.8 | 29.5 | 37.9 | 17.1 | 32.8 |
| | 35.8 | | 36.5 | 33.9 | 26.5 | 15.9 | 31.4 |
| | 35.9 | | 29.7 | 44.9 | 20.1 | 27.6 | 16.4 |
| | 51.3 | | 37.0 | 17.5 | 39.7 | 13.2 | 36.2 |
| | 50.0 | | 14.9 | 42.4 | 2.5 | 27.6 | 13.4 |
| | | | 45.6 | 6.4 | 36.2 | 23.6 | 31.0 |
| | | | 30.1 | 22.8 | 5.3 | 20.9 | 7.9 |
| | | | 24.6 | 31.6 | 27.9 | 24.8 | 36.9 |
| | | | 37.0 | 30.2 | 19.0 | 25.0 | 14.1 |
| | | | 44.4 | 19.7 | 21.0 | 25.2 | 6.7 |
| | | | 45.8 | 31.9 | 31.4 | 43.7 | 16.7 |
| | | | 30.6 | 14.8 | 19.7 | 23.7 | |
| | | | 42.5 | 32.3 | 20.9 | | |
| | | | 19.8 | 37.6 | 24.4 | | |
| | | | 32.1 | 26.7 | 34.8 | | |
| Averages, | | | | | | | |
| 58.2 | 44.2 | 35.8 | 33.7 | 27.8 | 25.4 | 21.1 | |

* From Herm. Ebbinghaus, 'Über das Gedächtniss,' pp. 93–103.

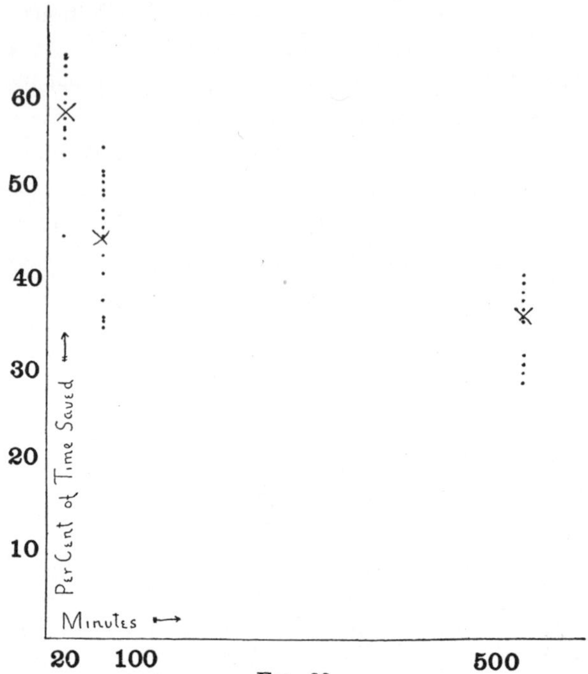


Fig. 82.

*Case III.*

If one mental trait is to be related to another the amounts of one are treated each as a constant and the problem is that of Case II., except for the fact that both series of amounts must be, unless there are real zero points, expressed as $k_1 + y_1$, $k_1 + y_2$, $k_1 + y_3$, etc., and $k_2 + x_1$, $k_2 + x_2$, etc.

The following case may be taken as an illustration:

The relationship between the ability to perceive *A*'s scattered among other capital letters and the ability to perceive words containing both *a* and *t* scattered among other words, the ability being measured in schoolgirls all of the 7*B* grammar grade. The amounts to be related are the number of *A*'s marked in 60 seconds and the number of words containing *a* and *t* marked in 120 seconds.

The related amounts found by measurement are given in Table XXXVI. The zero points being unknown, these pairs should all be turned into $k_1 + 10$ with $k_2 + 36$, $k_1 + 10$ with $k_2 + 51$, etc. If each related pair is plotted as before, our ignorance of the zero points would be expressed by leaving the axes of reference undetermined save in their direction, as in Fig. 83.

TABLE XXXVI.

| *a-t* words marked. | *A*'s marked. | *a-t* words marked. | *A*'s marked. | *a-t* words marked. | *A*'s marked. | *a-t* words marked. | *A*'s marked. |
|---|---|---|---|---|---|---|---|
| 10 | 36 | 17 | 47 | 20 | 58 | 23 | 62 |
| 10 | 51 | 17 | 49 | 20 | 60 | 23 | 65 |
| 11 | 43 | 17 | 57 | 20 | 61 | 23 | 70 |
| 11 | 47 | 18 | 41 | 20 | 62 | 24 | 55 |
| 11 | 56 | 18 | 43 | 20 | 64 | 24 | 55 |
| 12 | 45 | 18 | 46 | 20 | 76 | 24 | 59 |
| 12 | 46 | 18 | 47 | 21 | 45 | 24 | 78 |
| 13 | 52 | 18 | 47 | 21 | 46 | 25 | 49 |
| 13 | 55 | 18 | 51 | 21 | 47 | 25 | 54 |
| 14 | 48 | 18 | 51 | 21 | 48 | 25 | 59 |
| 14 | 58 | 18 | 53 | 21 | 49 | 25 | 70 |
| 15 | 37 | 18 | 62 | 21 | 50 | 25 | 78 |
| 15 | 38 | 18 | 62 | 21 | 54 | 25 | 81 |
| 15 | 42 | 18 | 63 | 21 | 54 | 26 | 57 |
| 15 | 43 | 18 | 66 | 21 | 57 | 26 | 60 |
| 15 | 47 | 19 | 57 | 21 | 59 | 27 | 61 |
| 15 | 50 | 19 | 60 | 21 | 59 | 27 | 64 |
| 15 | 52 | 19 | 61 | 21 | 61 | 27 | 65 |
| 15 | 64 | 19 | 64 | 21 | 63 | 27 | 67 |
| 15 | 72 | 20 | 38 | 21 | 65 | 27 | 74 |
| 16 | 43 | 20 | 43 | 22 | 47 | 27 | 78 |
| 16 | 46 | 20 | 45 | 22 | 48 | 28 | 54 |
| 16 | 46 | 20 | 46 | 22 | 53 | 28 | 65 |
| 16 | 55 | 20 | 48 | 22 | 59 | 28 | 65 |
| 16 | 56 | 20 | 50 | 22 | 62 | 29 | 69 |
| 16 | 67 | 20 | 51 | 22 | 62 | 30 | 49 |
| 16 | 70 | 20 | 52 | 22 | 63 | 30 | 59 |
| 17 | 39 | 20 | 56 | 22 | 77 | 30 | 81 |
| 17 | 42 | 20 | 56 | 23 | 45 | 34 | 73 |
| 17 | 44 | 20 | 56 | 23 | 48 | | |
| 17 | 45 | 20 | 57 | 23 | 58 | | |

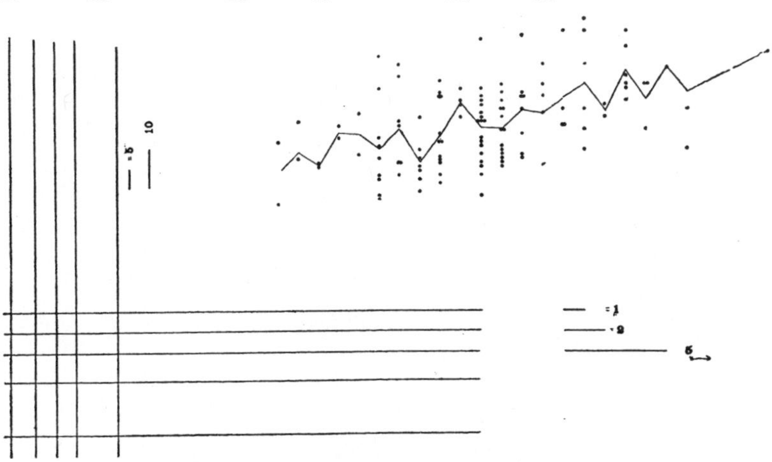

FIG. 83.

### *Case IV.*

The difficulty with zero points could be overcome if no attempt were made to measure the relations of absolute amounts, but only the relations of excesses or deficiencies similarly measured in a second trait. Thus, for instance, one may ask the relationship of the number of *A*'s marked in a minute more than 10 to the number of *a-t* words marked in two minutes more than 4; or the relationship of the number of *A*'s marked in a minute by ten-year-old boys more than the lowest record to a similar measure for *a-t* words; or a similar question with the average performance as the zero point in both cases. The last question is one that the mental sciences often ask; for the mental sciences are more frequently interested in the relationship of deviations in one trait from the general type to deviations in some other trait again from the general type, than in the relationship of gross amounts of the trait. The measurement now is simply of the

TABLE XXXVII.

| *a-t* words. | *A*'s | *a-t* words. | *A*'s | *a-t* words. | *A*'s | *a-t* words. | *A*'s | *a-t* words. | *A*'s |
|---|---|---|---|---|---|---|---|---|---|
| −10 | −19 | −3 | −16 | 0 | −7 | +1 | +10 | +7 | +6 |
| | −4 | | −13 | | −5 | +2 | −8 | | +9 |
| −9 | −12 | | −11 | | −4 | | −7 | | +10 |
| | −8 | | −10 | | −3 | | −2 | | +12 |
| | +1 | | −8 | | +1 | | +4 | | +19 |
| −8 | −10 | | −6 | | +1 | | +7 | | +23 |
| | −9 | | +2 | | +1 | | +7 | +8 | −1 |
| −7 | −3 | −2 | −14 | | +2 | | +8 | | +10 |
| | 0 | | −12 | | +3 | | +22 | | +10 |
| −6 | −7 | | −9 | | +5 | +3 | −10 | +9 | +14 |
| | +3 | | −8 | | +6 | | −7 | +10 | −6 |
| −5 | −18 | | −8 | | +7 | | +3 | | +4 |
| | −17 | | −4 | | +9 | | +7 | | +26 |
| | −13 | | −4 | | +21 | | +10 | +14 | +18 |
| | −12 | | −2 | +1 | −10 | | +15 | | |
| | −8 | | +7 | | −9 | +4 | 0 | | |
| | −5 | | +7 | | −8 | | 0 | | |
| | −3 | | +8 | | −7 | | +4 | | |
| | +9 | | +11 | | −6 | | +23 | | |
| | +17 | −1 | +2 | | −5 | +5 | −6 | | |
| −4 | −12 | | +5 | | −1 | | −1 | | |
| | −9 | | +6 | | −1 | | +4 | | |
| | −9 | | +9 | | +2 | | +15 | | |
| | 0 | 0 | −17 | | +4 | | +23 | | |
| | +1 | | −12 | | +4 | | +26 | | |
| | +12 | | −10 | | +6 | +6 | +2 | | |
| | +13 | | −9 | | +8 | | +5 | | |

relationship of the differences + or − of one trait from its typical condition to similar differences of the other trait.

If, after turning each of the measures of the previous illustration into terms of so much + or − the central tendency of the series to which it belongs, we treat them as in Case III., we have the results given in Table XXXVII. and Fig. 84.

These results represent in available form the relationship as found in each of the 122 cases studied. The variation among them is so great that any single law can express only the general tendency, a tendency from which the individuals often diverge very much.

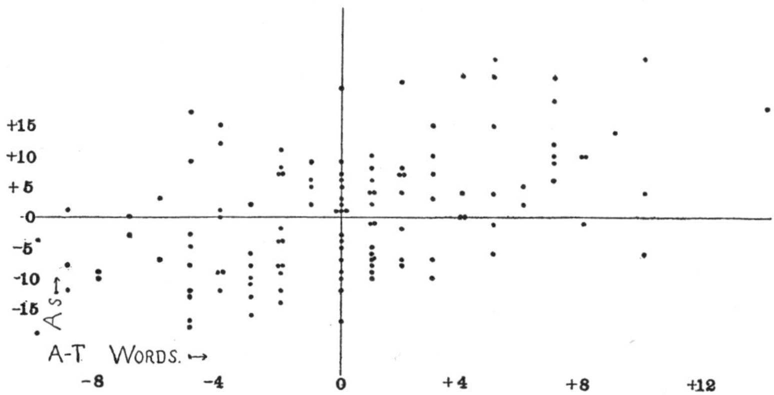


FIG. 84.

Each case of the relationship is, as has been shown, represented by the position of a point with reference to two axes or by an equation, *trait 1 = some function of trait 2, A = F of B.* Such tables and figures as XXXVI. and XXXVII. and 83 and 84, express together all the cases of a relationship which one has measured. The present problem is to find some simpler means of presenting the general tendency manifested by the total group of cases of the relationship.

The obviously useful habit of classifying the cases according to the amounts of one quantity to which the other is to be related has already been adopted. The group of measures in trait 2 related to any single measure in trait 1 is called the *array* correlated with that amount of 1. − 19 *A*'s and − 4 *A*'s form the array correlated with − 10 of *a-t* words; − 12, − 8 and + 1 form the array correlated with − 9 *a-t* words, etc. If in place of each array one takes its central tendency, the complicated table becomes the simple series of equations of Table XXXVIII.; the complicated diagram the simple series of points referred to two axes shown in Fig. 85. The problem is reduced to the same problem as in Case I., except for the difference in meaning of the axes of reference.

TABLE XXXVIII.

| Ability in *a-t* test. | Average of its array. | Number of cases in the array. | Ability in *a-t* test. | Average of its array. | Number of cases in the array. |
|---|---|---|---|---|---|
| —10 | —11.6 | 2 | + 1 | — .9 | 14 |
| — 9 | — 6.3 | 3 | + 2 | + 3.9 | 8 |
| — 8 | — 9.5 | 2 | + 3 | + 3.0 | 6 |
| — 7 | — 1.5 | 2 | + 4 | + 6.75 | 4 |
| — 6 | — 2.0 | 2 | + 5 | +10.2 | 6 |
| — 5 | — 5.55 | 9 | + 6 | + 3.5 | 2 |
| — 4 | — .6 | 7 | + 7 | +13.2 | 6 |
| — 3 | — 8.8 | 7 | + 8 | + 6.3 | 3 |
| — 2 | — 2.3 | 12 | + 9 | +14.0 | 1 |
| — 1 | + 5.5 | 4 | +10 | + 8.0 | 3 |
| 0 | — .6 | 18 | +14 | +18.0 | 1 |

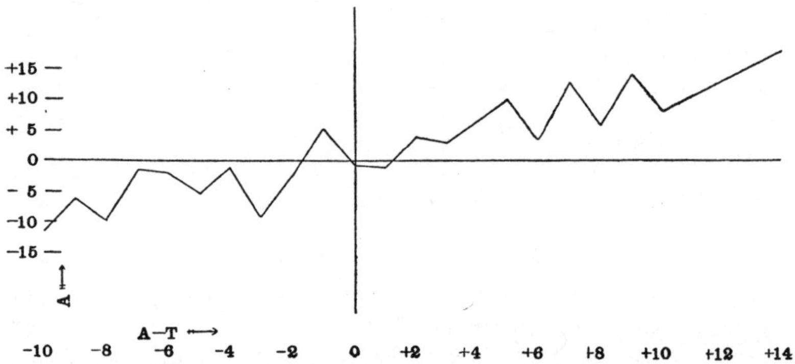


FIG. 85.

If the measure of the general tendency of each array were determined exactly, the general tendency of the relationship would be exactly determined by the equations of Table XXXVIII. and the position of the points in Fig. 85. We should have nothing to do but state them. For instance, suppose that in the *A a-t* relationship the results were :

A measure of —5 *a-t* words has a related array with a central tendency of —10.00
" —4 " " " " " — 8.00
" —3 " " " " " — 6.00
" —2 " " " " " — 4.00
" —1 " " " " " — 2.00

| | | | | | | | |
|---|---|---|---|---|---|---|---|
| A measure of | —0 *a-t* | words has a related | array | with a central | tendency | of — | 0 |
| " | +1 | " | " | " | " | " | + .50 |
| " | +2 | " | " | " | " | " | + 1.00 |
| " | +3 | " | " | " | " | " | + 1.50 |
| " | +4 | " | " | " | " | " | + 1.50 |
| " | +5 | " | " | " | " | " | + 1.50 |

These facts would be the general tendency of the relationship. One would simply say:

"Individuals who in the *a-t* test mark a given number less than the average will in the *A* test, on the whole, mark twice that number less than the average; individuals at the average in one will be at the average in the other. Individuals marking 1, 2 or 3 more than the average in the *a-t* test will mark one half that number more than the average in the *A* test. Individuals marking over + 3 in the *a-t* test do as well and no better in the *A* test than those marking + 3."

But, in fact, a relationship is almost never determined at all exactly for each particular amount of the first trait, especially not for the extreme + and – amounts. From relatively inexact measures of the general tendencies of the different arrays we infer the character of the relationship as a whole.

In making the inference the first step is to decide whether it is proper to assume that the general tendency of the relationships is uniform for all amounts of trait 1, is of the form $A = B$ times a constant, is such that the line through the points representing the exact or true central tendencies of the arrays is a straight line. In technical terms, can it be assumed that the correlation is rectilinear?

It is clear that even if the true correlation were thus rectilinear, the chance unreliabilities of the central tendencies of the actual arrays due to the small number of cases would make the relationship vary in amount for different arrays. It is also the fact that mental relationships apparently approximate to the rectilinear type more than to any one other. It is, therefore, customary to make the assumption unless there is some special reason for not doing so. The criteria which one might use to establish a warranted decision will be explained later. For the present we may best inquire what the next step in inference is, granted that the true relationship is of the form $A = B$ times a constant, that the line of correlation is rectilinear.

If for the true relationship $A/B =$ a constant, the ratio $A/B$ will be approximated by the central tendency of all the ratios actually found in individual cases and the true line of correlation will be approximated closely by the straight line from which the points as in Fig. 85 diverge least. In other words, if we find the central tendency of all the individual ratios (which are given in Table XXXIX.) or the straight line which fits best all the points of Fig. 85, we shall have an approximation to the true relationship.

TABLE XXXIX.

Ratios Expressing the Individual Relationships of Table XXXVII.

| | | | | |
|---|---|---|---|---|
| 1.90 | 5.33 | | 10.00 | .86 |
| .40 | 4.33 | | —4.00 | 1.29 |
| 1.33 | 3.67 | | —3.50 | 1.43 |
| .89 | 3.33 | | —1.00 | 1.71 |
| — .11 | 2.66 | | 2.00 | 2.71 |
| 1.25 | 2.00 | | 3.50 | 3.29 |
| 1.13 | — .67 | | 3.50 | —.13 |
| .43 | 7.00 | | 4.00 | 1.25 |
| 00 | 6.00 | | 11.00 | 1.25 |
| 1.17 | 4.50 | | —3.33 | 1.56 |
| — .50 | 4.00 | | —2.33 | —.60 |
| 3.60 | 4.00 | | 1.00 | .40 |
| 3.40 | 2.00 | | 2.33 | 2.60 |
| 2.60 | 2.00 | | 3.33 | 1.29 |
| 2.40 | 1.00 | —10.00 | 5.00 | |
| 1.60 | —3.50 | — 9.00 | 00 | |
| 1.00 | —3.50 | — 8.00 | 00 | |
| .60 | —4.00 | — 7.00 | 1.00 | |
| —1.80 | —5.50 | — 6.00 | 5.75 | |
| —3.40 | — 2.00 | — 5.00 | —1.20 | |
| 3.00 | —5.00 | — 1.00 | — .20 | |
| 2.25 | —6.00 | — 1.00 | .80 | |
| 2.25 | —9.00 | 2.00 | 3.00 | |
| 00 | | 4.00 | 4.75 | |
| — .25 | | 4.00 | 5.20 | |
| —3.00 | | 6.00 | .33 | |
| —3.25 | | 8.00 | .83 | |

Each of these methods has, however, a serious defect. In calculating the central ratio of the observed individual ratios, the ratios for any amount of one trait count as much as those for any other amount. The ratio 2.00 obtained from a case of — 1 *a-t* words with — 2 *A*'s plays as much of a rôle as the ratio 2.00 from a case of — 10 *a-t* words with — 20 *A*'s. But the latter should count more, since chance variation is far more likely to deflect an individual up or down 2 *A*'s than 20.

In calculating the straight line to best fit the series of points, a point ascertained from an array with few cases counts as much as a point ascertained from an array with many. The fifth point from the right, due to two cases, counts as much as the sixth point, which is due to nine cases. But, of course, the knowledge of a relationship's amount due to nine cases is much more reliable and deserving of weight than that due to two.

These difficulties would be removed if the second method could be so modified that the line drawn would be that from which the entire series of points of Fig. 84 diverged least, or in fitter terms, would be that expressing the general tendency of relationship from which all the individual relationships found would most probably result.

The Pearson method of calculating the general tendency of a relationship assumed to be rectilinear does this, and is, therefore, a method of the utmost service to the student of causal and other relationships in the mental sciences. The formula used and convenient ways of making the necessary calculations will be explained later.

The final desideratum in the measurement of a relationship is that it be intelligible in itself and commensurable with measurements of other relationships.

It is obviously misleading to say that a girl who is 14 above in the *a-t* test and 26 above in the *A* test is 186 per cent. as far above in the latter as in the former. In both cases she is the best girl of the group and is in reality, therefore, equally far above the average. Similarly, girls who were + 4 in the *a-t* and + 9 in the *A* test would really be equally superior in both, for they would be in both the 23d to 26th persons from the top out of the 122. Distance from the average in each case must, if the two cases are to be commensurate, be in terms of the variability of the distribution. The variabilities are: *a-t* test, A. D. = 3.57; *A* test, A. D. = 8.33. Case 1 on one list should really be scored − 10./3.57 and − 19/8.33, giving the ratio .82. The table of ratios thus corrected becomes Table XL.

The diagram may be corrected by dividing each measure by the variability of the distribution to which it belongs, or more easily by arranging the scale on the diagram so as to make the proper allowance and then using the original figures.

The difficulty in comparing different relationships, due to the

TABLE XL.

INDIVIDUAL RATIOS OF TABLE XXXIX. CORRECTED FOR THE VARIABILITY OF EACH TRAIT.

| | | | | |
|---|---|---|---|---|
| 82 | 229 | | 429 | 37 |
| 17 | 186 | | —172 | 55 |
| 57 | 157 | | —150 | 61 |
| 38 | 143 | | — 43 | 73 |
| — 5 | 114 | | 86 | 116 |
| 54 | 86 | | 150 | 141 |
| 48 | — 29 | | 150 | — 6 |
| 18 | 300 | | 172 | 54 |
| 00 | 257 | | 472 | 54 |
| 50 | 193 | | —143 | 67 |
| — 21 | 172 | | —100 | — 26 |
| 154 | 172 | | 43 | 17 |
| 146 | 86 | | 100 | 112 |
| 112 | 86 | | 143 | 55 |
| 103 | 43 | —429 | 215 | |
| 69 | —150 | —386 | 00 | |
| 43 | —150 | —343 | 00 | |
| 26 | —172 | —300 | 43 | |
| — 77 | —236 | —257 | 247 | |
| —146 | — 86 | —215 | — 51 | |
| 129 | —215 | — 43 | — 9 | |
| 97 | —257 | — 43 | 34 | |
| 97 | —386 | 86 | 129 | |
| 00 | | 172 | 204 | |
| — 11 | | 172 | 223 | |
| 129 | | 257 | 14 | |
| 139 | | 343 | 36 | |

fact that the units of measure for the different traits are incommensurate, disappears if they are each and all reduced to terms of the variability of the group. They then become commensurate, indeed identical, in the sense that in each of the tests the best person of 10,000 chosen at random would be plus the same figure. The 10th best in the one would be plus the same amount as the 10th best in any other,* etc.

The estimation of any relationship for a group would then be comparable with that of any other relationship for that group, and many now awkward questions of the mental and social sciences would be amenable to exact and readily obtained answers. The Pearson method of calculating rectilinear relationships fulfills this desideratum, and thus meets the exacting demands of a measure of the

* This would hold exactly only in so far as the forms of distribution of the different traits were alike.

general tendency of a relationship between two variable quantities with unknown zero points and units directly incommensurable.

The Pearson method obtains as its measure of the relationship a single number, which may be anywhere between 1.00 and − 1.00. A coefficient of correlation between two abilities of + 100 per cent. means that the individual who is the best in the group in one ability will be the best in the other, that the worst man in the one will be the worst in the other; that if the individuals were ranged in order of excellence in the first ability and then in order of excellence in the second, the two rankings would be identical; that any one's station in the one will be identical with his station in the other (both being reduced to terms of the variabilities of the abilities as units to allow comparison). A coefficient of − 100 per cent. would, per contra, mean that the best person in the one ability would be the worst in the other, that any degree of superiority in the one would go with an equal degree of inferiority in the other, and *vice versa.* A coefficient of + 62 per cent. would mean that (comparison being rendered fair here as always by reduction to the variabilities as units) any given station in the one trait would imply 62 hundredths of that station in the other. A coefficient of − 62 would, of course, mean that any degree of superiority would involve 62 hundredths as much inferiority, and *vice versa.*

The method of calculating the Pearson coefficient of correlation is to multiply each case's deviation from the average in the one trait by its deviation from the average in the other trait; to add together all the products thus found and divide their sum by the number of cases times the standard deviation of the first trait times the standard deviation of the second trait. That is, the coefficient of correlation,

$$r = \frac{\Sigma x \cdot y}{n\sigma_1\sigma_2}.$$

The arithmetic involved in calculating Pearson coefficients is simple, and, though lengthy, does not take so long a time as might be supposed. The apparently tedious process of multiplication can be done quickly and with no mental effort by the use of Crelle's Rechentafeln,* which is a multiplication table running to 1,000 times 1,000.

* Published by Georg Reimer, Berlin. The price is about $4.50. A multiplication table up to 100 times 100 is given in Appendix I. of this book.

The squaring involved in the calculation of $\sigma_1$ and $\sigma_2$ is, of course, done with the aid of a table of squares, such as Barlow's tables.* The addition is tedious unless one has at his service an adding machine. Even without an adding machine, however, a coefficient can be calculated from 1,000 individual relationships under the most unfavorable circumstances in less than a day. Different ways of arranging the material economize time in different cases. The procedure which is most generally serviceable is to calculate the average for each array and then replace $\Sigma xy$ by [(av. of first array of $B$) × (amount of $A$ with which it is correlated) × (its number of cases)] + [(av. of second array of $B$) × (amount of $A$ with which it is correlated) × (its number of cases)] etc., through the last array. This reduces the addition in part to multiplication and gives us knowledge of the degree to which the relationship approaches a rectilinear form. Thus in the case of our illustration we obtain the facts of Table XLI.

TABLE XLI.

| *A*<br>Various amounts of Trait 1. *a–t* words. | *B*<br>Averages of related arrays. | *C*<br>Number of cases in the arrays. | *D*<br>Averages of array times amount of Trait 1 times frequency, *i. e.*, *A* × *B* × *C*. |
|---|---|---|---|
| —10 | —11.50 | 2 | 230 |
| — 9 | — 6.33 | 3 | 171 |
| — 8 | — 9.50 | 2 | 152 |
| — 7 | — 1.50 | 2 | 21 |
| — 6 | — 2.00 | 2 | 24 |
| — 5 | — 5.56 | 9 | 250 |
| — 4 | — .57 | 7 | 16 |
| — 3 | — 8.86 | 7 | 186 |
| — 2 | — 2.33 | 12 | 56 |
| — 1 | + 5.50 | 4 | 22 |
| 0 | | 18 | 0 |
| + 1 | — .93 | 14 | 13 |
| + 2 | + 3.88 | 8 | 62 |
| + 3 | + 3.00 | 6 | 54 |
| + 4 | + 6.75 | 4 | 108 |
| + 5 | +10.17 | 6 | 305 |
| + 6 | + 3.50 | 2 | 42 |
| + 7 | +13.17 | 6 | 553 |
| + 8 | + 6.33 | 3 | 152 |
| + 9 | +14.00 | 1 | 126 |
| +10 | + 8.00 | 3 | 240 |
| +14 | +18.00 | 1 | 252 |
| | | | 2,965 |

* The squares and square roots of the numbers up to 1,000 are given in Appendix II. of this book.

The relationship may fairly be assumed to be rectilinear from the figures in column *B*, and their graphic representation in Fig. 85. Using the Pearson formula, then, we have

$$\Sigma xy = 2{,}965. \quad \sigma_1 = 4.65 \text{ (see calculation on page 126).}$$
$$\sigma_2 = 10.1 \quad \text{“} \quad \text{“} \quad \text{“} \quad \text{“}$$
$$n = 122$$

$r$, the coefficient of correlation, then equals $+ .52$. If for other reasons it is known to be valid to assume rectilinear correlation, it is somewhat quicker to calculate $\Sigma xy$ directly from the individual records. This calculation in full, together with that of $\sigma_1$ and $\sigma_2$ is given in Table XLII. Ordinary arithmetical skill could much abbreviate the calculation given there by combining multiplicands in multiplying and so saving later addition.

TABLE XLII.

*A.* — Calculation of $\Sigma xy$.

| | | | | | | | | | |
|---|---|---|---|---|---|---|---|---|---|
| 190 | | 48 | | 0 | | 10 | | 42 | |
| 40 | | 39 | | 0 | | | —16 | 63 | |
| 108 | | 33 | | 0 | | | —14 | 70 | |
| 72 | | 30 | | 0 | | | — 4 | 84 | |
| | — 9 | 24 | | 0 | | 8 | | 133 | |
| 80 | | 18 | | 0 | | 14 | | 161 | |
| 72 | | | — 6 | 0 | | 14 | | | — 8 |
| 21 | | 28 | | 0 | | 16 | | 80 | |
| 0 | | 24 | | 0 | | 44 | | 80 | |
| 42 | | 18 | | 0 | | | —30 | 126 | |
| | —18 | 16 | | 0 | | | —21 | | —60 |
| 90 | | 16 | | 0 | | 9 | | 40 | |
| 85 | | 8 | | 0 | | 21 | | 260 | |
| 65 | | 8 | | 0 | | 30 | | 252 | |
| 60 | | 4 | | | —10 | 45 | | | |
| 40 | | | —14 | | — 9 | 0 | | | |
| 25 | | | —14 | | — 8 | 0 | | | |
| 15 | | | —16 | | — 7 | 16 | | | |
| | —45 | | —22 | | — 6 | 92 | | | |
| | —85 | | — 2 | | — 5 | | —30 | | |
| 48 | | | — 5 | | — 1 | | — 5 | | |
| 36 | | | — 6 | | — 1 | 30 | | | |
| 36 | | | — 9 | 2 | | 75 | | | |
| 0 | | | | 4 | | 115 | | | |
| | — 4 | 0 | | 4 | | 130 | | | |
| | —48 | 0 | | 6 | | 12 | | | |
| | —52 | 0 | | 8 | | 30 | | | |

The sum of the $xy$ products $= 2{,}965$.

*B.* — Calculation of $\sigma_1$ and $\sigma_2$.

| | | |
|---|---|---|
| $-20^2 \times$ | 1 | $=400$ |
| $-19^2$ | 1 | 381 |
| $-18^2$ | 2 | 648 |
| $-17^2$ | 1 | 289 |
| $-15^2$ | 1 | 225 |
| $-14^2$ | 2 | 392 |
| $-13^2$ | 5 | 845 |
| $-12^2$ | 1 | 144 |
| $-11^2$ | 5 | 605 |
| $-10^2$ | 6 | 600 |
| $-9^2$ | 7 | 567 |
| $-8^2$ | 5 | 320 |
| $-7^2$ | 4 | 196 |
| $-6^2$ | 3 | 108 |
| $-5^2$ | 4 | 100 |
| $-4^2$ | 3 | 48 |
| $-3^2$ | 2 | 18 |
| $-2^2$ | 4 | 16 |
| $-1^2$ | 4 | 4 |
| 0 | 5 | 0 |
| $+1^2$ | 5 | 5 |
| $+2^2$ | 3 | 12 |
| $+3^2$ | 6 | 54 |
| $+4^2$ | 3 | 48 |
| $+5^3$ | 4 | 100 |
| $+6^2$ | 6 | 216 |
| $+7^2$ | 3 | 147 |
| $+8^2$ | 4 | 256 |
| $+9^2$ | 5 | 225 |
| $+10^2$ | 1 | 100 |
| $+11^2$ | 2 | 242 |
| $+13^2$ | 1 | 169 |
| $+14^2$ | 3 | 588 |
| $+16^2$ | 1 | 256 |
| $+17^2$ | 1 | 289 |
| $+18^2$ | 1 | 324 |
| $+20^2$ | 1 | 400 |
| $+21^2$ | 1 | 441 |
| $+22^2$ | 3 | 1,452 |
| $+25^2$ | 2 | 1,250 |
| | | 12,480 |

$12,480 \div 122 = 102.3$

$\sqrt{102.3} = 10.1$

$N\sigma_1\sigma_2 = 122 \times 10.1 \times 4.65$

$N\sigma_1\sigma_2 = 5730$

$r = 2965/5730, \quad r = +.52$

| | | |
|---|---|---|
| $-10^2 \times$ | 2 | $= 200$ |
| $-9^2$ | 3 | 243 |
| $-8^2$ | 2 | 128 |
| $-7^2$ | 2 | 98 |
| $-6^2$ | 2 | 72 |
| $-5^2$ | 9 | 225 |
| $-4^2$ | 7 | 112 |
| $-3^2$ | 7 | 63 |
| $-2^2$ | 12 | 48 |
| $-1^2$ | 4 | 4 |
| 0 | 18 | 0 |
| $+1^2$ | 14 | 14 |
| $+2^2$ | 8 | 32 |
| $+3^2$ | 6 | 54 |
| $+4^3$ | 4 | 64 |
| $+5^2$ | 6 | 150 |
| $+6^2$ | 2 | 72 |
| $+7^2$ | 6 | 294 |
| $+8^2$ | 3 | 192 |
| $+9^2$ | 1 | 81 |
| $+10^2$ | 3 | 300 |
| $+14^2$ | 1 | 196 |
| | | 2,642 |

$2,642 \div 122 = 21.656$

$\sqrt{21.656} = 4.65$

All the discussion of measurements of relationship so far presupposes that the facts related are measured exactly. There will, however, in mental and social measurements commonly be a considerable error in each individual fact of those to be related. For instance, in our illustration the '*A*'s marked by each individual' is a score depending upon only one trial of 60 seconds. With many trials on many different occasions, the individuals concerned would attain somewhat different measures. So also with the '*a-t* words marked.' Let us call $r_{\text{acc. m.}}$ the $r$ which would be obtained in our illustration from accurate measures in both traits for all of the individuals, and $r_{\text{app. m.}}$ the $r$ which is in fact calculated from the single measures. $r_{\text{acc. m.}}$ will be *greater* * than $r_{\text{app. m.}}$, for the influence of chance inaccuracy in the measures to be related is always to produce zero correlation. If two series of pairs of values are due entirely to chance the correlation will be zero, and in so far as they are at all due to chance, they will reduce the correlation.

The chance variation, which in the long run cuts its own throat in the case of averages and variabilities, can not in the case of a relationship be thus rendered innocuous by mere numbers. For instance the true relationship between the volume of bodies of water at constant pressure and temperature, etc., and their weight is + 1.00. Suppose now that the true measures for ten pairs were:

| Case. | Vol. | Wt. |
|---|---|---|
| *A* | 2 | 4 |
| *B* | 4 | 8 |
| *C* | 6 | 12 |
| *D* | 7 | 14 |
| *E* | 8 | 16 |
| *F* | 9 | 18 |
| *G* | 10 | 20 |
| *H* | 11 | 22 |
| *I* | 13 | 26 |
| *J* | 15 | 30 |

The correlation is evidently + 1.00.

Suppose the person measuring them got instead of these figures certain chance variations from them due to the error of his measuring.

If the reader will distribute by chance among these 20 errors, say 5 of 1, 5 of − 1, 4 of 2, 4 of − 2, 1 of 3 and 1 of − 3 and then

* By greater is meant more plus if the relationship from accurate measures is positive, more minus if it is negative.

calculate again the coefficient, he will find it to be less than before. If he will let the chance errors be larger, *e. g.*, 5 each of + 2 and − 2, 4 each of + 4 and − 4 and 1 each of + 6 and − 6, the coefficient will be still more reduced. The same will hold regardless of whether 10 or 10,000 pairs of related values are taken.

To correct for this 'attenuation' of the coefficient by chance errors in the data, it is necessary to have at least two independent measures of the measures to be related. When these are at hand the procedure is as follows:

Let $A$ and $B$ be the traits to be related.

Let $p$ be a series of exact measures of $A$.

Let $q$ be the related series of exact measures of $B$.

Denote by $r_{pq}$ the coefficient of correlation of $A$ and $B$, obtainable from the two series $p$ and $q$. $r_{pq}$ is thus the required real relationship.

Denote by $r_{p'q'}$ the average of the correlations between each series of values obtained for trait $A$ and each series of related values for trait $B$.

Denote by $r_{p'p''}$ the average of the correlations between any one series of measures of trait $A$ and any other corresponding series of independent measures of trait $A$.

Denote by $r_{q'q''}$ the average of the correlations between any one series of measures of trait $B$ and any other corresponding series of independent measures of trait $B$.

$$\text{Then } r_{pq} \doteq \frac{r_{p'q'}}{\sqrt{(r_{p'p''})(r_{q'q''})}}.$$

Thus if we have two series of independent measures of trait $A$ and similarly of the related trait $B$, if, that is, we have certain individuals measured twice in each trait, we shall have as our formula

$$r_{pq} = \frac{\dfrac{r_{p_1q_1} + r_{p_1q_2} + r_{p_2q_1} + r_{p_2q_2}}{4}}{\sqrt{(r_{p_1p_2})(r_{q_1q_2})}},$$

in which $p_1$ and $p_2$ refer to the two independent series of measures of trait $A$; $q_1$ and $q_2$ refer to the two independent series of measures of trait $B$; $r_{p_1p_2}$ is the coefficient of correlation between the first and second measures of $A$; $r_{q_1q_2}$ is the coefficient of correlation between the first and second measures of $B$; $r_{p_1q_1}$ is the coefficient of correlation between the first measure of $A$ and the first measure of $B$; $r_{p_1q_2}$ is the coefficient of correlation between the first measure of $A$ and the second measure of $B$; $r_{p_2q_1}$ is the coefficient of correlation between the second measure of $A$ and the first measure of $B$; $r_{p_2q_2}$ is the coefficient of correlation between the second measure of $A$ and the second measure of $B$.

A second method * of allowing for the inaccuracy of the original measures of the facts to be related is based upon the obvious fact that an increase in the number of measures of each of such facts increases its accuracy. From the increase in the closeness of the relationship as we use the central tendency of 2, 3, 4, 5 . . . trials of each individual, we may prophesy what the relationship would be if we had at hand measures from so many trials of all the individuals as to give the central tendencies exactly.

Let $r_{pq}$ be the coefficient of correlation that would be found if the measures of the related facts, $A$ and $B$, were perfectly exact.

Let $m$ be the number of independent measures of $A$, $p_1p_2p_3$, etc.

Let $n$ " " " " " " " $B$, $q_1q_2q_3$, etc.

Let $r_{p'q'}$ be the average of the correlations between each series of values obtained for trait $A$, with each series obtained for trait $B$.

Let $r_{p''q''}$ be the correlation obtained when $p_1p_2p_3$, etc. are combined to give the measure of trait $A$, when, that is, each individual is represented by his most likely central tendency in trait $A$, and when $q_1q_2q_3$ are similarly combined to give the measure of trait $B$.

$$\text{Then } r_{pq} = \frac{\sqrt[4]{mn}(r_{p''q''}) - r_{p'q'}}{\sqrt[4]{mn} - 1}.$$

Useful as these formulæ for correction of attenuation due to inaccurate measures are, it is wise not to overwork them by substituting their use for the attainment of reasonably precise original measures. The beginner, at all events, may best work here only with original measures, the P. E.$_{\text{true}-\text{obtained}}$ † of which is not over 5 per cent. of their amount.

Another source of error, a much less important one in practice,

* For a further description of this method and the first method as well see the article in the *Am. J. of Psy.*, for January, 1904, by C. Spearman, to whom the formulæ are due.

† See next chapter for the explanation of this term.

is the inaccuracy of the central tendencies from which the deviations are measured. The true relationship is of course that existing between the deviations of one series of measures from their true central tendency and the corresponding deviations of the second series from their true central tendency. The effect of inexact measures of the central tendencies is to make the obtained coefficient larger * than the true coefficient when the inaccuracies are both in the same direction and smaller when they are in different directions. The error is inconsiderable for inaccuracies such as occur in central tendencies calculated from 100 or more individual measures.

A third source of error deserves mention, though it is logical rather than statistical. To measure the relation between quality *A* and quality *B*, we should have a series of pairs of amounts related only through the relationship of *A* to *B*. But unless great care is taken in the selection of the data, other factors affecting the relationship of the amounts are sure to enter. Thus in relating mental capacities, if we use children of different ages, the factor of age, as well as the intrinsic relationship between the traits, is at work. The real relation between a city's lighting and its need of police protection might be inverse but actual correlations of the per capita expense for the two items in American cities might show a direct relationship due to the entrance of the factor, municipal expensiveness as a whole. The influence of heredity can not be inferred from fraternal correlation until a discount is made for the factor, similar training. Means of correcting for irrelevant factors have been devised, but it is safest to get data free from them in the first instance.

On page 119 the problem,' How to decide whether a relationship may be assumed to be rectilinear?' was suggested and postponed. It can not be given an absolute answer. One can, by knowing the unreliability of each array's central tendency, measure the likelihood that any given straight line chosen could be the true line of correlation. But some slightly crooked line would have a still greater likelihood. So far as the figures go, the most likely true relationship is the crooked line that passes through every point. It is because of a general confidence that nature is simple rather than complex, that regularity in relationships is more likely than irregularity, that we

* By larger is meant more plus in case the coefficient is positive, more minus in case it is negative ; by smaller is meant the reverse.

assume that the unevenness of the correlation found would disappear with more cases. If the student plots the line of central tendencies of arrays and on either side of it a line at the distance from it of the P. E.$_{\text{true central tendency} - \text{obtained central tendency}}$,* and then finds that the straight line which best fits the central tendency points falls in nine out of ten cases within the P. E. lines, he will rarely be wrong in assuming correlation to be rectilinear.

If correlation is demonstrably not rectilinear the mode of expressing its nature and amount will, of course, vary. The general problem will be, as always, to express the general tendency of relationship from which the actually found relationships can be derived with least improbability. Acquaintance with the concrete data concerned and natural ingenuity and insight will here be of far more service than cut and dried methods of technical procedure.

In presenting results no Pearson coefficient or other single expression should be given without also the total correlation table, or at least a diagram or list of the averages of the arrays such as may enable the reader to judge how far the relationship throughout is that expressed by the single ratio.

The facts to be related in the mental and social sciences may be either (1) the varying conditions of a trait in an individual (to be related to corresponding conditions in him of some other trait) or (2) the varying conditions of a trait found in different individuals of a group (to be related to the conditions found in some other trait in the same individuals) or (3) the varying central tendencies of a trait found in different subgroups of a larger group or collection of groups (to be related to the central tendencies found in the case of some other trait in the same subgroups).

For example, one may seek (Case 1) the relation between the quickness of perception of an individual at various times and his quickness of movement at corresponding times. Or one may seek (Case 2) the relation between the quickness of perception in general of Jones, Smith, Brown, etc., and the quickness of movement possessed in general by the same individuals. Or (Case 3) one may seek the relationship between the general quickness in perception of races to their quickness of movement.

* The meaning of this quantity may be left undefined until the next chapter is read.

It should be noted that the difference in the three cases is not in the mere number of individuals studied. The essential difference would remain if we used a million cases to determine the relationship of two traits within an individual, only a hundred thousand to determine the relationship among individuals and only ten thousand to determine it for races. The essential difference is in the questions to be solved. From them it follows also that in Case 1 if several individuals are studied a number of pairs of figures for each individual will be used and the general tendency of the relationship in each individual will be worked out separately. If the results from different individuals are then combined they will be combined as a group of facts according to the methods of Chapter IV. In Case 2, on the contrary, a single pair of figures will represent the relationship in any one individual and these pairs will be combined according to the method of the present chapter. In Case 3 a single pair of figures will represent the relationship in each subgroup.

The problem of measurement itself is the same for three cases, the difference being in the data used and the consequent meaning of the coefficient of correlation obtained. To any one of the following series of related pairs the mode of procedure discussed in this chapter is applicable.

RELATED BY IDENTITY OF CONDITIONS.

Trait $T$ and trait $T_1$ in individual $A$ under conditions $C_1$
" " " $C_2$
" " " $C_3$

RELATED BY IDENTITY OF THE INDIVIDUAL.

Trait $T$ and trait $T_1$ in group, ten-year-olds, in individual $I_1$
" " " $I_2$
" " " $I_3$

RELATED BY IDENTITY OF THE SUBGROUP.

Trait $T$ and trait $T_1$ in group, all men, in subgroup Chinese.
" " " Negroes.
" " " Indians.

It is perhaps needless to point out that the existence of a certain relationship within an individual does not imply anything about the relationship within a group of individuals, nor that again about the relationship within a group of groups. Individuals may be happier when they are richer, but rich individuals amongst Americans may be no happier than poor individuals, and from neither fact could we infer that the American population would be happier or less happy than the Chinese or the Negro population.

For similar reasons the nature and amount of a relationship will depend upon the group selected. If, for instance, the relationship between knowledge of history and knowledge of English literature is measured in the group, high-school graduates, by using the deviations of individuals from the high-school graduates' averages in the two traits, the relationship will be less close than if we use the group, all people. The relationship between height and weight will be less close if measured in the group, 18-year-olds, than if measured in all children under twenty. Any relationship so calculated should always be thought of as the relationship of deviations from the averages in the two traits in the individuals of *the group in question.* To assume that the relationship found in any given group holds good also for a different group is valid only if the given group is a random selection from the other group.

### *Application of the Theory of Measurements of Variable Relationships to the Problem of Measuring Mental Inheritance.*

The measurement of mental inheritance involves the measurement of similarities between related individuals and the measurement of the amount of such similarity to be attributed tò training. The first problem is statistically identical with that of measuring the relationship between two mental traits, only here the two traits will be the same trait in two related individuals, and the coefficient of correlation will measure not the implication of one trait with respect to another in the same man, but the implication of one trait in one man with respect to the same trait in his relative. In the formula, that is, the $xy$ products will be each the product of one person's deviation and that of his relative; $\sigma_1$ will be the variability of all the first members of the series of related pairs and $\sigma_2$ the variability of all the second members. $N$ will be the number of pairs.

### *Application to the Study of Causal Relationships.*

The possibility of measuring relationships conveniently and precisely is one step toward the study of causes in the mental sciences. It gives us a means of making Mill's method of 'concomitant variations' exact and applicable to variable facts. It allows us to make use of the criterion that the cause must be equal to the effect. Whenever one finds two quantities correlated he may properly proceed to test the hypotheses that one causes the other in part and that both are due in part to some common cause.

The point of view of this long chapter may be summed up in a few short practical precepts. They are:

Think what you are relating, and that any relationship is measured by a series of ratios.

If the measures are absolute amounts, bear in mind the significance of the zero points from which they are measured.

If the measures are deviations from some central tendency, bear in mind the nature of the group whose central tendency it is.

Keep before you always the total series of ratios found.

Do not be satisfied with crude means of measuring any presumably rectilinear relationship. The Pearson coefficient requires not much more time and is, for both exactness and convenience, far superior.

## Problems.

29. Calculate the relationship between changes in pauperism and changes in out-relief from the following data: *

Percentage Ratios of Pauperism.

| Percentage Ratios of Out-Relief Ratio. | 15–25 | 25–35 | 35–45 | 45–55 | 55–65 | 65–75 | 75–85 | 85–95 | 95–105 | 105–115 | 115–125 |
|---|---|---|---|---|---|---|---|---|---|---|---|
| 15–25 | | | 1 | | | | | | | | |
| 25–35 | 1 | 1 | | 4 | | | 1 | | | | |
| 35–45 | | 3 | 2 | 10 | 3 | 3 | | | | | |
| 45–55 | | 2 | 4 | 7 | 8 | 6 | 4 | | | | 1 |
| 55–65 | | | 4 | 10 | 11 | 11 | 8 | | | | |
| 65–75 | | | | 4 | 10 | 13 | 7 | 2 | 1 | | |
| 75–85 | | | 1 | 7 | 12 | 8 | 1 | 7 | 1 | | |
| 85–95 | | | | 1 | 4 | 3 | 1 | 1 | | 1 | |
| 95–105 | | | | 1 | 4 | 5 | 4 | 5 | | | |
| 105–115 | | | | | 1 | 4 | 5 | 1 | | | |
| 115–125 | | | | | 1 | 3 | 1 | 3 | 1 | 1 | |
| 125–135 | | | | | 1 | | 1 | 1 | 1 | | |
| 135–145 | | | | | | | 1 | | | | |
| 145–155 | | | | | | | | | | | |
| 155–165 | | | | | | | | | | | |
| 165–175 | | | | | | | | | | | |
| 175–185 | | | | | | | | | | | |
| 185–195 | | | | | | 1 | | | | | |

* From an article by G. Udny Yule, in the *Journal of the Royal Statistical Society*, Vol. 62, p. 281.

Each figure in the table represents the number of cases of the relationship denoted by the figure above it in the horizontal scale taken with the figure opposite it in the vertical scale. Thus the second column reads: 'Of districts having a change of 25–35 in pauperism, one had a change of 25–35 in out-relief ratio, three had changes of 35–45 in out-relief ratio, and 2 had changes of 45–55.

# CHAPTER X.

## THE RELIABILITY OF MEASURES.

WHEN from a limited number of measurements of an individual fact, say of *A*'s monthly expenses or *B*'s ability in perception, we calculate its average, the result is not, except by chance, the true average. For, obviously, one more measurement will, unless it happens to coincide with the average obtained, change it. For instance, the first 30 measures of *H*'s ability in reaction time gave the average .1405; the next seven measures being taken into account, the average became .1400; with the next seven it became .1406 —; with the next seven, .1406 +. By the true average we mean the average that would come from all the possible tests of the trait in question. The actual average obtained from a limited finite number of these measures is, except by chance, only an approximation toward the true average. So also with the accuracy of the measure of variability obtained. The true variability is that manifested in the entire series of measurements of the trait; the actually obtained variability is an approximation toward it. The true average and the true variability of a group mean similarly the measures obtained from a study of all the members of the group.

It is necessary, then, to know how many trials of an individual, how many members of a group, must be measured, to obtain as accurate knowledge as we need. Or, to speak more properly, it is necessary to know how close to the true measure the result obtained from a certain finite number of measures will be.

It is clear that the true average of any set of measures is the average calculated from all of them. If the average we actually obtain is calculated from samples chosen at random, it will probably diverge somewhat from the average calculated from all. So also with obtained and true measures of total distribution, variability, of difference and of relationship. We measure the unreliability of any obtained measure by its probable divergence from the true measure.

It is clear also that the divergence of any measure due to a limited number of measures from the corresponding measure due to the entire series, will vary according to what particular samples we hit upon, and that if the samples are taken at random this variation in the amount of divergence will follow the laws of probability. For these laws, based on the algebraic law expressing the number of combinations of $r$ things taken $n$ at a time, will account for the difference between the constitution of a total series and the constitution of any group of things chosen at random from it, consequently for the differences between any two measures due respectively to these two constitutions.

We have, consequently, to find the distribution of a *divergence* (of obtained from true or of true from obtained) and know beforehand, in cases of random sampling, that it will be of the type of the probability surfaces given in Figs. 12 and 49, will be symmetrical (since the true is as likely to be greater as to be less than the obtained) with its mode at 0 (since all that we do know about the true is that it is more likely to be the obtained measure than to be any other one measure). What we need to know is its form and variability, to know, that is, how often we may expect a divergence of .01, how often one of .02, how often one of .03, etc. Suppose our obtained measure to be 10.4 and the distribution of the probable divergence of its corresponding true measure from it to be known to be as follows:

| | | |
|---|---|---|
| —1.1 to — .9 | 1 or | .01 per cent. |
| — .9 " — .7 | 10 " | 1 " |
| — .7 " — .5 | 45 " | 4.5 " |
| — .5 " — .3 | 120 " | 12 " |
| — .3 " — .1 | 210 " | 21 " |
| — .1 " + .1 | 252 " | 25 " |
| + .1 " + .3 | 210 | |
| + .3 " + .5 | 120 | |
| + .5 " + .7 | 45 | |
| + .7 " + .9 | 10 | |
| + .9 " +1.1 | 1 | |

We can say: 'The true measure will not rise above 11.3 (10.4 + .9) in more than one case out of 1,024,' or, 'The chances are over 1,000 to 1 against the measure being over 11.3,' or, 'The chances are nearly 99 to 1 against the true measure being over 11.1,' or, 'The chances are about 8 to 1 against the true measure differing from 10.4 either above or below by more than .5.'*

* It may appear strange to talk about the true measure, which is a fixed value, 'rising above' or 'being over,' but if the reader will bear in mind that we do not know just where it is fixed, but do know the probability of its being at this or that point, he will not misunderstand the terms used. They could not well be avoided without much circumlocution.

If the form of the distribution of the divergence were known, its variability would be the only measure needed. The form will always be fairly near to the normal surface of frequency and it is customary to disregard the very slight error involved and assume the form to be normal.

If we know the variability of the divergence, the probable frequency of any divergence or of divergences less than or greater than any given amount can be calculated from the table of frequencies of the normal probability surface. Conversely, the table will tell us the amount of divergence which will be exceeded (or not exceeded) by any given per cent. of comparisons of true and obtained. Illustrations of the use of the table will be given in Chapter XI.

The problem of determining the reliability of any measure due to a limited series of samples is, then, to determine the variability of the fact, *divergence of true from obtained measure.* ($M_{\text{true}} - M_{\text{obt.}}$.)

It is clear that the more nearly the number of samples taken approaches the number of things they represent the closer the obtained measure will, in general, be to the true measure, the less will be the range of divergence.

It is clear that the less the variability amongst the individual samples, the less will be the divergence of the obtained from the true measure of central tendency. For instance, if men range from 4 to 7 feet in height, averaging 5 feet 8 inches, we can not possibly get an average more than 1 foot 8 inches wrong, while if they range from 2 to 10 feet, we may make an error of 3 feet 8 inches. The same holds true for the divergence of obtained from true variability.

Upon these facts are based the formulæ for the calculation of the variability of the divergence of true measure from that obtained from any given series of samples. These formulæ take as the definition of 'true measure,' the measure which would be found if an infinite number of cases were studied.

The formulæ to be given here for the reliability of central tendencies and variabilities are those in common use. They are absolutely exact only for a case where the distribution of the trait itself is that of the normal probability surface with extremes at minus infinity and plus infinity, and so are never absolutely exact for any real case. They are very inexact, except for a trait showing a clear central tendency with decreasing frequencies on either side. This, however, commonly occurs in those mental measurements from which we have any right, according to the principles of Chapters III. and IV., to calculate a type and divergences from it. The A. D. and $\sigma$ formulæ give in such cases a variability for the divergence of true from obtained that is a trifle too large, and so make the obtained result seem less reliable than it is. This is perhaps a useful error.

*The Reliability of an Average.*

The probable divergence of the true from the obtained average, depending upon the number of cases and the variability of the distribution, may be calculated according to different formulæ, according as we use $\sigma_{\text{dis.}}$, A. D.$_{\text{dis.}}$ or P. E.$_{\text{dis.}}$* as a measure of the variability of the distribution from which the average was obtained.

If we use $\sigma_{\text{dis.}}$, the divergence of the true from the obtained average will be a quantity symmetrically distributed about 0 as its mode or average, with a variability expressed by a mean square deviation of $\sigma_{\text{dis.}}/\sqrt{n}$. That is, $\sigma_{\text{t. av.}-\text{obt. av.}} = \sigma_{\text{dis.}}/\sqrt{n}$.

Its variability in terms of A. D. will be $.7979\sigma_{\text{dis.}}/\sqrt{n}$. That is, A. D.$_{\text{t. av.}-\text{obt. av.}} = .7979\sigma_{\text{dis.}}/\sqrt{n}$.

Its variability in terms of P. E. will be $.6745\sigma_{\text{dis.}}/\sqrt{n}$. That is, P. E.$_{\text{t. av.}-\text{obt. av.}} = .6745\sigma_{\text{dis.}}\sqrt{n}$.

For instance, let $A_{\text{obt.}}$ = the obtained average: let $\sigma_{\text{dis.}}$ = the variability (mean square or standard deviation) of the distribution: let $A_{\text{t.}}$ = the average that would be obtained from an infinite number of measures. Then, if $A_{\text{obt.}} = 20.2$, $\sigma_{\text{dis.}}$ 4.2 and the number of measures, 300, $A_{\text{t.}} - A_{\text{obt.}} = 0$ with $\sigma_{t-0}$ equal to 4.2/17.32 or .242, $A_{\text{t.}} - A_{\text{obt.}}$ will then range between $-.726$ and $+.726$ in 997 cases out of 1,000, between $-.242$ and $+.242$ in 682 cases out of 1,000, between $-.40$ and $+.40$ in 900 cases out of 1,000. The student can verify these figures from the table on page 148. In other words, the chances are 997 to 3 or 332 to 1, that the true average will not deviate from the obtained by more than .726; 682 to 318, or over 2 to 1, against a deviation of over .242; and 900 to 100, or 9 to 1, against a deviation of over .40. In still different words, the chances

*Since to measure reliability we have to measure the variability of a divergence and shall need to use terms similar to those used in measuring the variability of individual things or conditions, it will be well to name the average deviation of a distribution of a thing or condition A. D.$_{\text{dis.}}$. Similarly, $\sigma$ and P. E. in the sense hitherto used will now be called $\sigma_{\text{dis.}}$ and P. E.$_{\text{dis.}}$.

are 2 to 1 that the true average lies between 19.958 and 20.442; 9 to 1 that the true average lies between 19.8 and 20.6; 332 to 1 that the true average lies between 19.474 and 20.926.

If for a measure of the original distribution's variability we take its A. D.$_{\text{dis.}}$, the variability of the divergence of true from obtained average will be

$$\sigma_{\text{t. av.}-\text{obt. av.}} = \frac{1.2533 \text{ A. D.}_{\text{dis.}}}{\sqrt{n}}$$

$$\text{A. D.}_{\text{t. av.}-\text{obt. av.}} = \frac{\text{A. D.}_{\text{dis.}}}{\sqrt{n}}$$

$$\text{P. E.}_{\text{t. av.}-\text{obt. av.}} = \frac{.84435 \text{ A. D.}_{\text{dis.}}}{\sqrt{n}}$$

If for the measure of the original distribution's variability we take its P. E.$_{\text{dis.}}$, the variability of the divergence of true from obtained average will be

$$\sigma_{\text{t. av.}-\text{obt. av.}} = \frac{1.4826 \text{ P. E.}_{\text{dis.}}}{\sqrt{n}}$$

$$\text{A. D.}_{\text{t. av.}-\text{obt. av.}} = \frac{1.1843 \text{ P. E.}_{\text{dis.}}}{\sqrt{n}}$$

$$\text{P. E.}_{\text{t. av.}-\text{obt. av.}} = \frac{\text{P. E.}_{\text{dis.}}}{\sqrt{n}}$$

The same formulæ may be used roughly for the reliability of a median if the student himself remembers and warns his readers that the divergence of true from obtained median may exceed the amount shown by the formulæ. Actually the excess is not enough to lead to serious error.

For the mode too the same formulæ may be used as a rough approximation. In proportion as the mode is taken to cover a relatively wide unit the formulæ will give too great apparent unreliability. But in proportion as the mode is assumed on the mere basis of greatest frequency they will give the reverse.

This process of finding the probable divergence of true from obtained measure may be better realized by testing it experimentally. For example, let us take as *T*'s true average in some trait the average from 1,000 trials, and suppose the 1,000 trials to be distributed as follows:

| Quantity. | Frequency. | Quantity. | Frequency. |
|---|---|---|---|
| 10 | 10 | 17 | 140 |
| 11 | 20 | 18 | 130 |
| 12 | 40 | 19 | 100 |
| 13 | 80 | 20 | 80 |
| 14 | 100 | 21 | 50 |
| 15 | 120 | 22 | 20 |
| 16 | 110 | | |

*T*'s true average is then 16.51.

If now one takes 1,000 discs or slips of paper and marks 10 of them 10, 20 of them 11, 40 of them 12, etc., he can imitate the action of random selection in tests by random drawings from the discs. If one draws, say 20, and, regarding each as one trial in the tests, computes the average of the 20, he has the parallel of an obtained average from $N = 20$. The patience to make 100 or so drawings of 20 or so each will be rewarded by the opportunity to distribute the 100 obtained divergences of $\text{Av.}_{\text{true}}$ from $\text{Av.}_{\text{obt.}}$ and to see how far this distribution conforms to that obtained from the formulæ above.

In a similar experiment, where $\text{Av.}_{\text{true}}$ was 2.5 and 27 drawings, each of 20 from a series of 400, were made, the actual divergences were as given in column I. below. The probable divergences given by the average of the 27 formulæ,

$$\text{P. E.}_{\text{t. av.}-\text{av. obt. from first 20}} = \frac{\text{P. E.}_{\text{dis. of first 20}}}{\sqrt{20}},$$

$$\text{P. E.}_{\text{t. av.}-\text{av. obt. from second 20}} = \frac{\text{P. E.}_{\text{dis. of second 20}}}{\sqrt{20}},$$

etc., are given in column II. The $\text{P. E.}_{\text{t.}-\text{obt.}}$ from experiment is .212; that from theory is .224.

$A_{\text{t.}} - A_{\text{obt.}}$

| | By experiment. | By theory. | | By experiment. | By theory. |
|---|---|---|---|---|---|
| —.6 and beyond | 0 | .9+ | 0 to +.1 | 3 | 3.2 |
| —.5 to —.6 | 1 | .9— | +.1 " +.2 | 3 | 3.0— |
| —.4 " —.5 | 2 | 1.2+ | +.2 " +.3 | 5 | 2.4+ |
| —.3 " —.4 | 1 | 1.9+ | +.3 " +.4 | 1 | 1.9+ |
| —.2 " —.3 | 4 | 2.4+ | +.4 " +.5 | 0 | 1.2+ |
| —.1 " —.2 | 3 | 3.0— | +.5 " +.6 | 1 | .9— |
| 0 " —.1 | 3 | 3.2 | +.6 and beyond | 0 | .9+ |

*The Reliability of a Measure of Variability.*

As before, we are measuring a variable fact, 'Divergence of true from obtained variability,' which has a mode at 0, the distribution of a probability surface, and a variability calculated from the original series' variability and number of cases.

The formulæ for these measures of variability are for deviations from the average, but they may be used approximately for deviations from the mode or median.

The variability of the divergence of the true variability from the obtained variability is found from the following formulæ:

$$\sigma_{\text{t. var.}-\text{obt. var.}} = \frac{\sigma_{\text{dis.}}}{\sqrt{2n}} \text{ or } \frac{1.2533 \text{ A. D.}_{\text{dis.}}}{\sqrt{2n}} \text{ or } \frac{1.4826 \text{ P. E.}_{\text{dis.}}}{\sqrt{2n}},$$

$$\text{A. D.}_{\text{t. var.}-\text{obt. var.}} = \frac{.7979\sigma_{\text{dis.}}}{\sqrt{2n}} \text{ or } \frac{\text{A. D.}_{\text{dis.}}}{\sqrt{2n}} \text{ or } \frac{1.1843 \text{ P. E.}_{\text{dis.}}}{\sqrt{2n}},$$

$$\text{P. E.}_{\text{t. var.}-\text{obt. var.}} = \frac{.6745\sigma_{\text{dis.}}}{\sqrt{2n}} \text{ or } \frac{.84435 \text{ A. D.}_{\text{dis}}}{\sqrt{2n}} \text{ or } \frac{\text{P. E.}_{\text{dis.}}}{\sqrt{2n}}.$$

*The Reliability of a Measure of Difference.*

The unreliability of a difference, say between $A_{\text{obt.}}$ and $B_{\text{obt.}}$, is measured by means of the variability of the divergence between the two measures. The probable true measure $A_{\text{t.}}$ is distributed about $A_{\text{obt.}}$ as a mode and the probable true measure $B_{\text{t.}}$ is distributed about $B_{\text{obt.}}$ as its mode. The probable true difference, that is, $A_{\text{t.}} - B_{\text{t.}}$, is a variable with its mode at $A_{\text{obt.}} - B_{\text{obt.}}$ and with decreasing frequencies as we take $A_{\text{obt.}} - B_{\text{obt.}} + 1$, $A_{\text{obt.}} - B_{\text{obt.}} + 2$, etc., or $A_{\text{obt.}} - B_{\text{obt.}} - 1$, $A_{\text{obt.}} - B_{\text{obt.}} - 2$, etc. This may be seen most clearly in a concrete case such as follows:

Given the facts that $A_{\text{obt.}} = 42$ and $B_{\text{obt.}} = 50$, that the differences between $A_{\text{true}}$ and $A_{\text{obt.}}$ are as given in I., and the differences between $B_{\text{true}}$ and $B_{\text{obt.}}$ are as given in II.

| | I. | II. |
|---|---|---|
| Difference. | Frequency. | Frequency. |
| —2 to —3 | 1 | 1 |
| —1 " —2 | 5 | 5 |
| 0 " —1 | 10 | 10 |
| 0 " +1 | 10 | 10 |
| +1 " +2 | 5 | 5 |
| +2 " +3 | 1 | 1 |

To find the difference between $A_{true}$ and $B_{true}$. From I. and II. we get as probable values of $A_{true}$ and $B_{true}$, III. and IV.

| III. | | IV. | |
|---|---|---|---|
| $A_{true}$. | | $B_{true}$. | |
| 40 to 39 | 1 | 48 to 47 | 1 |
| 41 " 40 | 5 | 49 " 48 | 5 |
| 42 " 41 | 10 | 50 " 49 | 10 |
| 42 " 43 | 10 | 50 " 51 | 10 |
| 43 " 44 | 5 | 51 " 52 | 5 |
| 44 " 45 | 1 | 52 " 53 | 1 |

Using for each distance its midpoint value, $A_{true}$ and $B_{true}$ are:

| $A_{true}$. | | $B_{true}$. | |
|---|---|---|---|
| 39.5 | 1 | 47.5 | 1 |
| 40.5 | 5 | 48.5 | 5 |
| 41.5 | 10 | 49.5 | 10 |
| 42.5 | 10 | 50.5 | 10 |
| 43.5 | 5 | 51.5 | 5 |
| 44.5 | 1 | 52.5 | 1 |

From these probable values of $A_{true}$ and $B_{true}$ we get the following probable differences between $A_{true}$ and $B_{true}$:

| | | | | | | | |
|---|---|---|---|---|---|---|---|
| One | 39.5 | with one 47.5 | gives | 1 | difference | of | 8 |
| | | " five 48.5s | " | 5 | differences | " | 9 |
| | | " ten 49.5s | " | 10 | " | " | 10 |
| | | " ten 50.5s | " | 10 | " | " | 11 |
| | | " five 51.5s | " | 5 | " | " | 12 |
| | | " one 52.5 | " | 1 | difference | " | 13 |
| Five | 40.5s | with one 47.5 | give | 5 | differences | " | 7 |
| | | " five 48.5s | " | 25 | " | " | 8 |
| | | " ten 49.5s | " | 50 | " | " | 9 |
| | | " ten 50.5s | " | 50 | " | " | 10 |
| | | " five 51.5s | " | 25 | " | " | 11 |
| | | " one 52.5 | " | 5 | " | " | 12 |
| Ten | 41.5s | with one 47.5 | " | 10 | differences | " | 6 |
| | | " five 48.5s | " | 50 | " | " | 7 |
| | | " ten 49.5s | " | 100 | " | " | 8 |
| | | " ten 50.5s | " | 100 | " | " | 9 |
| | | " five 51.5s | " | 50 | " | " | 10 |
| | | " one 52.5 | " | 10 | " | " | 12 |
| Ten | 42.5s | with one 47.5 | " | 10 | differences | " | 5 |
| | | " five 48.5s | " | 50 | " | " | 6 |
| | | " ten 49.5s | " | 100 | " | " | 7 |
| | | " ten 50.5s | " | 100 | " | " | 8 |
| | | " five 51.5s | " | 50 | " | " | 9 |
| | | " one 52.5 | " | 10 | " | " | 10 |

| | | | | | |
|---|---|---|---|---|---|
| Five 43.5s | with one 47.5 | gives | 5 differences | of | 4 |
| | " five 48.5s | " | 25 " | " | 5 |
| | " ten 49.5s | " | 50 " | " | 6 |
| | " ten 50.5s | " | 50 " | " | 7 |
| | " five 51.5s | " | 25 " | " | 8 |
| | " one 52.5 | " | 5 " | " | 9 |
| One 44.5 | with one 47.5 | " | 1 difference | " | 3 |
| | " five 48.5s | " | 5 differences | " | 4 |
| | " ten 49.5s | " | 10 " | " | 5 |
| | " ten 50.5s | " | 10 " | " | 6 |
| | " five 51.5s | " | 5 " | " | 7 |
| | " one 52.5 | " | 1 difference | " | 8 |

Putting together all these differences between $A_{\text{true}}$ and $B_{\text{true}}$, we have:

| Frequency. | | | | | | | | Quantity. |
|---|---|---|---|---|---|---|---|---|
| 1 | probable | difference | between | $A_{\text{true}}$ | and | $B_{\text{true}}$ | of | 3 |
| 10 | " | differences | " | " | " | " | " | 4 |
| 45 | " | " | " | " | " | " | " | 5 |
| 120 | " | " | " | " | " | " | " | 6 |
| 210 | " | " | " | " | " | " | " | 7 |
| 252 | " | " | " | " | " | " | " | 8 |
| 210 | " | " | " | " | " | " | " | 9 |
| 120 | " | " | " | " | " | " | " | 10 |
| 45 | " | " | " | " | " | " | " | 11 |
| 10 | " | " | " | " | " | " | " | 12 |
| 1 | " | difference | " | " | " | " | " | 13 |

This table is the distribution of $A_{\text{true}} - B_{\text{true}}$. The mode is $-8$ ($A$ being less than $B$) or $A_{\text{obt.}} - B_{\text{obt.}}$ The variability is P. E. = 1.12. Since the distribution of $A_{\text{true}} - B_{\text{true}}$ about $-8$ as a mode is the same thing as the distribution of the divergence of $A_{\text{true}} - B_{\text{true}}$ from $A_{\text{obt.}} - B_{\text{obt.}}$ about 0 as a mode, we have

$$\text{P. E.}_{(A_{\text{t.}} - B_{\text{t.}}) - (A_{\text{obt.}} - B_{\text{obt.}})} = 1.12.$$

The chances are 1 to 1 that the true difference will not vary from the obtained difference by more than 1.12, will not go outside of $-6.88$ and $-9.12$.

The variability of the divergence between the true measures is thus dependent on the variabilities of the divergences of each one from its corresponding obtained measure. The unreliability of a difference between two measures equals in fact the square root of the sum of the squares of the unreliabilities of the measures themselves. The formula for its calculation is, Variability of $(A_{\text{t.}} - B_{\text{t.}})$

$$= \sqrt{[\text{var. of } (A_{\text{t.}} - A_{\text{obt.}})]^2 + [\text{var. of } (B_{\text{t.}} - B_{\text{obt.}})]^2}.$$

Using the common standards of measurement of variability,

$$\sigma_{\text{diff. } A_{\text{t.}} - B_{\text{t.}}} = \sqrt{(\sigma_{A_{\text{t.}} - A_{\text{obt.}}})^2 + (\sigma_{B_{\text{t.}} - B_{\text{obt.}}})^2}$$

$$\text{A. D.}_{\text{diff. } A_{\text{t.}} - B_{\text{t.}}} = \sqrt{(\text{A. D.}_{A_{\text{t.}} - A_{\text{obt.}}})^2 + (\text{A. D.}_{B_{\text{t.}} - B_{\text{obt.}}})^2}$$

$$\text{P. E.}_{\text{diff. } A_{\text{t.}} - B_{\text{t.}}} = \sqrt{(\text{P. E.}_{A_{\text{t.}} - A_{\text{obt.}}})^2 + (\text{P. E.}_{B_{\text{t.}} - B_{\text{obt.}}})^2}$$

The most probable true difference is, then, the obtained difference, and the chances that the true difference is so much less or so much more than it can be calculated from the tables for the probability surface.

*The Reliability of a Pearson Coefficient of Correlation.*

The divergence, in a case of lineal correlation, of the true coefficient of correlation from that obtained from the limited number of pairs of measures compared, is a variable trait with a probable mode at 0, and a variability which serves as the measure of the unreliability of the obtained result. The formulæ * are:

$$\sigma_{r_{\text{t.}} - r_{\text{obt.}}} = \frac{1 - r^2}{\sqrt{n(1 + r^2)}}$$

$$\text{A. D.}_{r_{\text{t.}} - r_{\text{obt.}}} = \frac{.7979(1 - r^2)}{\sqrt{n(1 + r^2)}}$$

$$\text{P. E.}_{r_{\text{t.}} - r_{\text{obt.}}} = \frac{.6745(1 - r^2)}{\sqrt{n(1 + r^2)}}$$

It is customary to speak of the variability of the divergence of true from obtained measure as the measure's *error*. Thus $\sigma_{\text{t. av.} - \text{obt. av.}}$ is called the *mean square error* of the obtained average; $\text{P. E.}_{\text{t. r.} - \text{obt. r.}}$ is called the *probable error* of the obtained coefficient of correlation; $\text{A. D.}_{\text{t. diff.} - \text{obt. diff.}}$ is called the *average error* of the obtained difference. These terms are somewhat ill chosen, as there is really no 'error,' but only a varying degree of probable approximation. I have, therefore, used the word unreliability throughout.

## PROBLEMS.

What is the unreliability of each of the averages and variabilities in the following cases?

* There is some uncertainty about these formulæ, certain authorities favoring the use of simply $\sqrt{n}$ in the denominators in place of $\sqrt{n(1 + r^2)}$.

30. $Av._A = 10$. $P. E._{dis.} = 1$. $N = 20$.
31. $Av._B = 10$. " " 1.5. " " 30.
32. $Av._C = 12$. " " 2.0. " " 40.
33. $Av._D = 13$. " " 3.0. " " 40.
34. $Av._E = 14$. " " 3.0. " " 360.

What is the unreliability of each of the following differences?

35. $Av._C - Av._A = 2$. The data concerning $Av._A$ and $Av._C$ being as in 30 and 32.

36. $Av._D - Av._A = 3$. The data concerning $Av._A$ and $Av._C$ being as in 30 and 33.

37. $Av._E - Av._A = 4$. The data concerning $Av._A$ and $Av._C$ being as in 34 and 30.

38. $Av._E - Av._B = 4$. The data concerning $Av._A$ and $Av._C$ being as in 34 and 31.

39. $Av._E - Av._C = 2$. The data concerning $Av._A$ and $Av._C$ being as in 34 and 32.

What is the unreliability of $r$ in each of the following cases?

40. $r = .46$. $N = 200$.
41. $r = .16$. $N = 200$.
42. $r = .16$. $N = 600$.

## CHAPTER XI.

### THE USE OF TABLES OF FREQUENCY OF THE PROBABILITY SURFACE.

TABLE XLIII. gives for any normal surface of frequency the per cent. of cases included between the average, 0, and any degree of deviation, the latter being measured in terms of the standard deviation of the distribution, $\sigma_{dis.}$. Tables XLIV. and XLV. give the same information when the degrees of deviation are in terms of the A. $D._{dis.}$ and P. $E._{dis.}$.

Thus the first line of entries of Table XLIII. reads: Between the average and .01 $\sigma$ either above or below, either + or —, there are .004 of the cases; between the average and + .02$\sigma$ there are .008 of the cases; between the average and — .03$\sigma$ there are .0120 of the cases, etc.

It thus enables one to calculate the entire distribution of any trait which is normally distributed, the average and variability of which are known. For instance, if one finds for discrimination of color that the average = 24.0 and the standard deviation = 4.0, one finds from the table that the ability 24 — 24.99 or that between the average and + .25$\sigma$, will be possessed by 9.87 per cent. of the group; the ability 24 — 25.99 or that between 0 and + .5$\sigma$ by 19.15 per cent., and consequently the ability 25 — 25.99 by 19.15 — 9.87, or 9.28 per cent. By thus finding the percentages included between the average ability and different amounts of deviation from it, and so between any two given limits of deviation from it, one gets, as the table of frequencies in our illustrative case, Table XLVI.

This use of the tables gives a convenient means of measuring the degree to which the measures under investigation approximate to the probability curve distribution. If the table of actual frequencies of the measures is compared entry for entry with the frequencies given for corresponding deviations in the table for the probability curve, one can see at a glance the general closeness of correspondence. In making such comparisons the actual frequencies may properly be grouped so as to represent only 18 or more grades, and any most likely central point may be chosen with which to make the central point of the probability surface coincide.

For example, let the measures in the first column of frequencies of Table XLVII. be the actual distribution. Their average is 76 — ; their median, 76 + ; and their most likely mode, 75 — or 77 —. 76 may be taken as the central point for the comparison. Their A. D.$_{\text{dis.}}$ from it is 2.65 steps (5.30 units). In the second column the actual frequencies are given in per cents. From Table XLIV.

TABLE XLIII.

TABLE OF VALUES OF THE NORMAL PROBABILITY INTEGRAL CORRESPONDING TO VALUES OF $x/\sigma$ OR THE FRACTION OF THE AREA OF THE CURVE BETWEEN THE LIMITS 0 AND $+x/\sigma$ OR 0 AND $-x/\sigma$.

*Total area of curve assumed to be* 10,000.

$x$ = deviation from mean.

$\sigma$ = standard deviation.

| $x/\sigma$ | 0 | 1 | 2 | 3 | 4 | 5 | 6 | 7 | 8 | 9 | Δ |
|---|---|---|---|---|---|---|---|---|---|---|---|
| 0.0 | 0000 | 0040 | 0080 | 0120 | 0160 | 0200 | 0239 | 0279 | 0319 | 0359 | 40 |
| 0.1 | 0399 | 0438 | 0478 | 0517 | 0557 | 0597 | 0636 | 0676 | 0715 | 0754 | 40 |
| 0.2 | 0793 | 0832 | 0871 | 0910 | 0948 | 0987 | 1026 | 1064 | 1103 | 1141 | 39 |
| 0.3 | 1179 | 1217 | 1255 | 1293 | 1330 | 1368 | 1406 | 1443 | 1480 | 1517 | 38 |
| 0.4 | 1554 | 1591 | 1628 | 1664 | 1700 | 1737 | 1773 | 1808 | 1844 | 1879 | 36 |
| 0.5 | 1915 | 1950 | 1985 | 2020 | 2054 | 2089 | 2124 | 2157 | 2191 | 2225 | 34 |
| 0.6 | 2258 | 2291 | 2324 | 2357 | 2389 | 2422 | 2454 | 2486 | 2518 | 2549 | 32 |
| 0.7 | 2581 | 2612 | 2643 | 2672 | 2704 | 2734 | 2764 | 2794 | 2823 | 2853 | 30 |
| 0.8 | 2882 | 2910 | 2939 | 2967 | 2995 | 3023 | 3051 | 3078 | 3106 | 3133 | 28 |
| 0.9 | 3160 | 3186 | 3212 | 3238 | 3264 | 3290 | 3315 | 3340 | 3365 | 3389 | 26 |
| 1.0 | 3414 | 3438 | 3461 | 3485 | 3509 | 3532 | 3555 | 3577 | 3600 | 3622 | 23 |
| 1.1 | 3644 | 3665 | 3686 | 3708 | 3729 | 3750 | 3770 | 3791 | 3811 | 3830 | 21 |
| 1.2 | 3850 | 3869 | 3888 | 3906 | 3925 | 3944 | 3962 | 3980 | 3997 | 4015 | 19 |
| 1.3 | 4032 | 4049 | 4066 | 4083 | 4099 | 4115 | 4132 | 4147 | 4162 | 4178 | 17 |
| 1.4 | 4193 | 4208 | 4222 | 4237 | 4251 | 4265 | 4279 | 4292 | 4306 | 4319 | 14 |
| 1.5 | 4332 | 4345 | 4358 | 4370 | 4383 | 4395 | 4406 | 4418 | 4429 | 4441 | 12 |
| 1.6 | 4452 | 4463 | 4474 | 4485 | 4496 | 4506 | 4516 | 4526 | 4536 | 4545 | 10 |
| 1.7 | 4554 | 4564 | 4573 | 4582 | 4591 | 4600 | 4608 | 4617 | 4625 | 4633 | 9 |
| 1.8 | 4641 | 4648 | 4656 | 4664 | 4671 | 4678 | 4686 | 4693 | 4700 | 4706 | 7 |
| 1.9 | 4713 | 4720 | 4726 | 4732 | 4738 | 4744 | 4750 | 4756 | 4762 | 4767 | 6 |
| 2.0 | 4773 | 4778 | 4783 | 4788 | 4794 | 4799 | 4804 | 4808 | 4813 | 4817 | 5 |
| 2.1 | 4822 | 4826 | 4830 | 4834 | 4838 | 4842 | 4846 | 4850 | 4854 | 4858 | 4 |
| 2.2 | 4861 | 4865 | 4868 | 4872 | 4875 | 4878 | 4881 | 4884 | 4887 | 4890 | 3 |
| 2.3 | 4893 | 4896 | 4899 | 4901 | 4904 | 4906 | 4909 | 4911 | 4914 | 4916 | 3 |
| 2.4 | 4918 | 4921 | 4923 | 4925 | 4927 | 4929 | 4931 | 4933 | 4935 | 4936 | 2 |
| 2.5 | 4938 | 4940 | 4942 | 4943 | 4945 | 4946 | 4947 | 4949 | 4951 | 4952 | 2 |
| 2.6 | 4953 | 4955 | 4956 | 4958 | 4959 | 4960 | 4961 | 4962 | 4964 | 4965 | 1 |
| 2.7 | 4966 | 4967 | 4968 | 4969 | 4970 | 4970 | 4971 | 4972 | 4973 | 4974 | 1 |
| 2.8 | 4975 | 4975 | 4976 | 4977 | 4978 | 4978 | 4979 | 4980 | 4981 | 4981 | 0.5 |
| 2.9 | 4982 | 4982 | 4983 | 4983 | 4984 | 4984 | 4985 | 4985 | 4986 | 4986 | 0.5 |
| 3 | 4987 | 4991 | 4993 | 4995 | 4997 | 4998 | 4999 | 4999 | 4999 | 5000 | ...... |
| 8 | 5000 | ...... | ...... | ...... | ...... | ...... | ...... | ...... | ...... | ...... | ...... |

TABLE XLIV.

Table of Values of the Normal Probability Integral Corresponding to Values of $x/(\text{A. D.})$. Total Area of the Surface of Frequency Taken as 1,000.

| X/A. D. Multiples of the A. D. | .0 | .1 | .2 | .3 | .4 | .5 | .6 | .7 | .8 | .9 |
|---|---|---|---|---|---|---|---|---|---|---|
| 0. | 000 | 032 | 063 | 095 | 125 | 155 | 184 | 212 | 238 | 264 |
| 1. | 288 | 310 | 331 | 350 | 368 | 384 | 399 | 413 | 425 | 435 |
| 2. | 445 | 453 | 460 | 467 | 472 | 477 | 481 | 484 | 487 | 490 |
| 3. | 492 | 493.4 | 494.6 | 495.8 | 496.7 | 497.4 | 498.0 | 498.4 | 498.7 | 499.1 |
| 4. | 499.3 | 499.5 | 499.6 | 499.7 | 499.8 | 499.9 | | | | 500.0 |

TABLE XLV.

Table of Values of the Probability Integral Corresponding to Values of $X/(\text{P. E.})$. Total Area of the Surface of Frequency Taken as 1,000.

| X/P. E. Multiples of the P. E. | .0 | .1 | .2 | .3 | .4 | .5 | .6 | .7 | .8 | .9 |
|---|---|---|---|---|---|---|---|---|---|---|
| 0. | 000 | 027 | 054 | 080 | 106 | 132 | 157 | 182 | 205 | 228 |
| 1. | 250 | 271 | 291 | 310 | 328 | 344 | 360 | 374 | 388 | 400 |
| 2. | 411 | 422 | 431 | 440 | 447 | 454 | 460 | 466 | 471 | 475 |
| 3. | 479 | 482 | 485 | 487 | 489 | 491 | 493 | 494 | 495 | 496 |
| 4. | 497 | | 498 | | 499 | | | | | |
| 5. | 499.7 | | | | | | | | | |

TABLE XLVI.

Ability. Frequency in Per Cents., Av. Being 24.0, and $\sigma$ Being 4.0.

| Ability. | Frequency. | Ability. | Frequency. | Ability. | Frequency. |
|---|---|---|---|---|---|
| Less than 11 | 0.06 | 20–20.99 | 6.80 | 29–29.99 | 3.88 |
| 11–11.99 | 0.07 | 21 | 8.19 | 30 | 2.68 |
| 12 | 0.17 | 22 | 9.28 | 31 | 1.73 |
| 13 | 0.32 | 23 | 9.87 | 32 | 1.05 |
| 14 | 0.60 | 24 | 9.87 | 33 | 0.60 |
| 15 | 1.05 | 25 | 9.28 | 34 | 0.32 |
| 16 | 1.73 | 26 | 8.19 | 35 | 0.17 |
| 17 | 2.68 | 27 | 6.80 | 36 | 0.07 |
| 18 | 3.88 | 28 | 6.30 | 37 and over | 0.06 |
| 19 | 6.30 | | | | |

are calculated the frequencies of deviations $-.5/2.65$ to $+.5/2.65$, $+.5/2.65$ to $+1.5/2.65$, and so on. These are given in the third column of frequencies of Table XLVII. The divergences of the actual distribution from the probability curve distribution of the same central point and variability are given in the next column, and in the last column are put in per cents. of the corresponding probability surface frequencies. Fig. 41, on page 50, gives the comparison in terms of space.

TABLE XLVII.

Actual Distribution of Ratio of Attendance to Enrollment in Cities of U. S. Compared with Normal Distribution.

| Quantity. | I. Actual Frequency. | II. Actual Frequency in per cents. | III. Frequency in normal surface. | IV. Differences. | V. Differences in per cents., IV. is of III. |
|---|---|---|---|---|---|
| 45–46.9 | 1 | .184 | .025 (45–51) | + .184 | ? large |
| 7 | 1 | .184 | | + .184 | ? large |
| 9 | 0 | | | — .05 | ? large |
| 51 | 2 | .37 | | + .35 | ? large |
| 3 | 0 | | .055 | — .055 | — 100 |
| 5 | 4 | .74 | .12 | + .62 | + 52 |
| 7 | 1 | .184 | .34 | — .156 | — 46 |
| 9 | 2 | .37 | .66 | — .29 | — 44 |
| 61 | 4 | .74 | 1.3 | — .56 | — 43 |
| 3 | 15 | 2.75 | 2.4 | + .35 | + 15 |
| 5 | 21 | 3.85 | 3.8 | + .05 | + 1 |
| 7 | 34 | 6.24 | 5.9 | + .34 | + 6 |
| 9 | 44 | 8.07 | 7.8 | + .27 | + 3 |
| 71 | 31 | 5.69 | 10.1 | —4.32 | — 43 |
| 3 | 54 | 9.91 | 11.45 | —1.54 | — 13 |
| 5 | 65 | 11.9 | 11.9 | .000 | 0 |
| 7 | 89 | 16.3 | 11.45 | +4.85 | + 42 |
| 9 | 70 | 12.85 | 10.1 | +2.75 | + 27 |
| 81 | 37 | 6.79 | 7.8 | —1.01 | — 13 |
| 3 | 29 | 5.32 | 5.9 | — .58 | — 10 |
| 5 | 15 | 2.75 | 3.8 | —1.05 | — 28 |
| 7 | 11 | 2.03 | 2.4 | — .37 | — 15 |
| 9 | 9 | 1.65 | 1.3 | + .35 | + 27 |
| 91 | 1 | .184 | .66 | — .376 | — 57 |
| 3 | 2 | .37 | .34 | + .03 | + 9 |
| 5 | 1 | .184 | .12 | + .064 | + 53 |
| 7–98.9 | 2 | .37 | .055 | + .315 | + 573 |
| | | | .025 | | ? large |
| Total N = 545 | | | | | ? large |

To find the frequency of any given ability in a normal distribution, the central point and variability of which are known.

The frequency of any degree of ability can obviously be calculated quickly if the average and variability are given. For instance, if $A = 10$ and $\sigma = 2.4$, how many cases will be between 12.4 and 12.6? 12.4 is exactly $1\sigma$ from the Av. and 12.6 is $1.0833\sigma$ from the Av. The per cents. of cases included between $A$ and $1\sigma$ and between $A$ and $1.08\sigma$ are respectively 34.14 and 36.00. The number of cases between $1\sigma$ and $1.08\sigma$ is then 1.86 per cent. of the whole number in the series. To be exact and allow for the .0033, we add to the last figure one third of the difference in the table between the per cents. for 1.08 and 1.09, *viz.*, one third of a 22 or .0007. .3414 from .3607 then gives us .0193, or 1.93 per cent. The number of cases between 12.4 and 12.6 is, then, 1.93 per cent. of the whole number of cases. Practice with the following problems will familarize one with this use of the table:

43. Av. = 10. $\sigma = 3$. What per cent. of cases lie between 7 and 13?

44. Av. = 22. $\sigma = 4.4$. What per cent. of cases lie between 18 and 20?

45. Av. = 15.5. $\sigma = 2.1$. What per cent. of cases lie above 22?

46. Av. = 15.5. $\sigma = 2.1$. What per cent. of cases lie below 13?

47. Av. = 14.86. A. D. = 3.46. What per cent. of cases lie between 12 and 13?

48. Av. = 14.86. A. D. = 3.46. What per cent. of cases lie between 14 and 16?

49. Av. = 29.74. P. E. = 3.18. What per cent. of cases lie between 24 and 25?

To find, from any starting-point on the scale of measurement, the limits of ability that will include a stated percentage of the cases.

By using the tables the other way about, one may find, Av. and $\sigma$ being known, the degree of deviation from the average (or the distance from any stated point, *e. g.*, the upper limit, the lower limit, the point $1\sigma$ above, etc.) needed to include any stated percentage of the cases.

For instance, how far above the average must one go to get one fourth of the cases, the Av. being 8.0 and $\sigma$ 2.0? A distance of $.67\sigma$ includes 2,486 and a distance of $.68\sigma$ 2,518. A distance of $.675\sigma$ will obviously include 25 per cent., .675 times 2 is 1.35. Hence the answer is 9.35. Again, what limits of ability will include 80 per cent. of the cases? From knowledge of the shape of the normal surface it is known that the cases are thickest the nearer they are to the average. So, of course, we take in the example, limits equidistant from the average. They are $+ 1.28\sigma$ and $- 1.28\sigma$, or more exactly, $+ 1.2817\sigma$ and $- 1.2817\sigma$. In the illustration these are

5.4366 and 10.5634. In reckoning inward from either extreme it is best to arbitrarily take $3\sigma$ as the limit plus or minus, though in the theoretical surface the limits are plus infinity and minus infinity.

The following are simple problems:

50. Av. = 10 and $\sigma = 2$. What limits will include the 30 per cent. just above the average?

51. The 20 per cent. below it?

52. The middle two thirds of the cases?

53. Av. = 17.24. A. D. = 4.6. What limits will include the middle three fourths of the cases?

54. The bottom 10 per cent.?

55. The second sixth of the cases from the top?

This use of the tables is that followed in transmuting a series of measures in terms of relative position into terms of amount. In so far as the distribution of the trait is that of the probability surface we can, calling the average 0, find the limits of deviation from it in terms of the variability as a unit which will include, say, the lowest 1 per cent., the next 3 per cent., the 8 per cent. from the 23d to 31st per cent. from the top, etc. The process is so far identical with that in the examples just given. Then follows the calculation of an average amount to fit the cases included between each pair of limits. How this is done may be seen from a concrete case. Suppose that of 400 boys' themes 16, or 4 per cent., are indistinguishable for excellence, but are worse than 100 and better than 284. They are then per cents., 25, 26, 27 and 28. By Table *A* these per cents. will lie between $+ .6745\sigma$ and $+ .5531\sigma$. By the table we find that the abilities between these limits have the following frequencies:

| Ability. | Frequency. |
|---|---|
| $.5531\sigma$ to $.56\sigma$ | 23 |
| .56 | 34 |
| .57 | 34 |
| .58 | 34 |
| .59 | 33 |
| .60 | 33 |
| .61 | 33 |
| .62 | 33 |
| .63 | 32 |
| .64 | 33 |
| .65 | 32 |
| .66 | 32 |
| $.67\sigma$ to $.6745\sigma$ | 14 |

The average ability for the group is .61 +. This was the method by which Tables XXXI. and XXXII. in Chapter VII. were constructed.

Given the unreliability of an average in the form of the variability of its divergence from the true average ($\sigma_{\text{t. Av.}-\text{obt. Av.}}$ or A. D.$_{\text{t. Av.}-\text{obt. Av.}}$ or P. E.$_{\text{t. Av.}-\text{obt. Av.}}$); to calculate the chances that the true average will differ from the obtained by any given amount. The problem is simply that of finding the frequency of any degree of ability in a normal distribution the central point and variability of which are known.

For example, $\sigma_{\text{t. Av.}-\text{obt. Av.}}$ is 3.2. To find the chances that the true average will not vary from $A_{\text{obt.}}$ by more than 1.0, 2.0, 3.0, 4.0, 6.0 and 10.0. 1.0 is + 31 per cent. of 3.2. By the table deviations within the limits $+ .31\sigma$ and $-.31\sigma$ occur with a frequency of 12.17 + 12.17 or 24.34 per cent. There is, then, 1 chance out of 4 that $A_{\text{t.}}$ will not differ from $A_{\text{obt.}}$ by more than 1.0. 2.0 is 62 per cent. of 3.2. By the table deviations within the limits $+ .62\sigma$ and $- .62\sigma$ occur in 45.8 per cent. of the cases. The chances are almost 1 to 1 that $A_{\text{t.}}$ will not differ from $A_{\text{obt.}}$ by more than 2.0. The chances of a difference of less than 10 will be found to be 9,986 out of 10,000, or over 700 to 1.

Given the unreliability of $A_{\text{obt.}}$ in the same way as above, to calculate the amount of divergence of $A_{\text{t.}}$ from $A_{\text{obt.}}$ more than which has a given degree of improbability.

This problem, the converse of the above, is identical with that of calculating limits of ability from the average as a starting-point.

For example, $\sigma_{\text{t. av.}-\text{obt. av.}}$ is 3.0. To find the amount of difference between $A_{\text{t.}}$ and $A_{\text{obt.}}$, differences greater than which will have only 1 chance in 100 of happening. In the table we find the distance from the average which must be passed over in both plus and minus directions to include 99 out of 100 cases, 49.5 plus and 49.5 minus. It is $2.575\sigma$. Since $\sigma$ equals 3.0 the answer to our problem is 7.725.

It will be noted that the tables serve equally well in the many cases where the desired fact is the probability of a given divergence

Av. = 10 *in one direction* or the amount of divergence in one direction, more divergence than which has a given degree of improbability.

The same methods serve if the unreliability is of a variability or of a difference or of a relationship — in short, for all cases where the unreliability is measured by the variability of a divergence of true from obtained, and this divergence is distributed in a normal probability surface.

The following problems will offer opportunity for acquiring self-confidence in the use of the tables in connection with all sorts of questions about unreliability :

56. $\sigma_{t.-o.\ Av.} = 1.6$. (*a*) What is the probability of a difference between $Av._{t.}$ and $Av._{o.}$ of 4.0 or more? (*b*) What are the chances that $Av._{t.}$ will be 3.2 greater than $Av._{o.}$? (*c*) Between what limits will the true average lie with a probability of 9999 to 1?

57. $\sigma_{t.-o.\ var.} = .4$. (*a*) What is the probability that the true variability is more than .8 less than the obtained? (*b*) That the true variability is not more than .6 above or below the obtained?

58. $\sigma_{t.-o.\ diff.} = .5$. The actually obtained difference is, $Av._1 - Av._2 = 1.2$. (*a*) What is the probability that the true difference is zero or less than zero? (*b*) That the true difference is: $Av._1 - Av._2 = 2.4$ or more? (*c*) That the true superiority of $Av._1$ over $Av._2$ is between 1.7 and .7? (*d*) What limits would you assign for the true difference to be sure that the chances would be 20 to 1 against their being exceeded?

59. $r_{o.} = +.48$. $\sigma_{t.-o.\ rel.} = .04$. (*a*) Between what limits does the true relationship lie with practical certainty (it is customary to take 997 out of 1,000 as practical certainty)? (*b*) What is the chance that the true relationship is as low as .40?

60. $Av._{o.} = 22.6$. $A. D._{t.-o.\ Av.} = .4$. (*a*) What is the chance that the true average is as large as 24.0? (*b*) That it is as small as 22.0?

61. $Av._{o.} = 28.2$. $P. E._{t.-o.\ Av.} = .6$. (*a*) What is the chance that the true average is less than 26.0? (*b*) That it varies from $Av._{o.}$ by less than 2.0?

62. If it were true that the chances were 82 to 18 that the true average would not vary from the obtained by more than 13.4, what would be the value of $P. E._{t.-o.\ Av.}$?

63. $\text{Av.}_1 = 10.1$, $\text{Av.}_2 = 12.4$. $\text{P. E.}_{\text{t.}-\text{o. diff. of Av.}_1 \text{ and Av.}_2} = 1.0$. (*a*) What are the chances that $\text{Av.}_1 - \text{Av.}_2 = 0$ or less? (*b*) 1.0 or less? (*c*) 2.5 or more? (*d*) Between 2.0 and 2.8? (*e*) Between 1.0 and 3.3?

64. $\text{P. E.}_{\text{dis. obt.}} = 1.6$, $\text{A. D.}_{\text{t.}-\text{o. var.}} = 0.1$. (*a*) What are the chances that $\text{P. E.}_{\text{dis.}}$ will be between 1.4 and 1.8? (*b*) That it will not exceed 1.9? (*c*) What limits must be taken such that the true $\text{P. E.}_{\text{dis.}}$ will be practically certain (see question 59) not to exceed them?

65. $r_{\text{o.}} = +.39$, $\text{P. E.}_{\text{t.}-\text{o. rel.}} = .008$. What is the chance of the true relationship being as high as + 40? As + 41? As + .42? As + .50?

66. Speaking roughly, the true measure is practically certain to lie between the following limits:

Obtained measure $+ 3\sigma_{\text{t.}-\text{o. measure}}$ and obtained measure $- 3\sigma_{\text{t.}-\text{o. measure}}$.
" " $+ 3\frac{3}{4}\ \text{A. D.}_{\text{t.}-\text{o. measure}}$ and obtained measure $- 3\frac{3}{4}\ \text{A. D.}_{\text{t.}-\text{o. measure}}$.
" " $+ 4\frac{1}{2}\ \text{P. E.}_{\text{t.}-\text{o. measure}}$ " " " $- 4\frac{1}{2}\ \text{P. E.}_{\text{t.}-\text{o. measure}}$.

Justify this statement from the tables.

67. $r_{1_{\text{o.}}} - r_{2_{\text{o.}}} = .04$, $\text{P. E.}_{\text{t.}-\text{o. diff. } r_1 \text{ and } r_2} = .06$. (*a*) What is the chance that the true $r_2$ is really equal to or greater than the true $r_1$? (*b*) What is the chance that the true $r_1$ is greater than the true $r_2$?

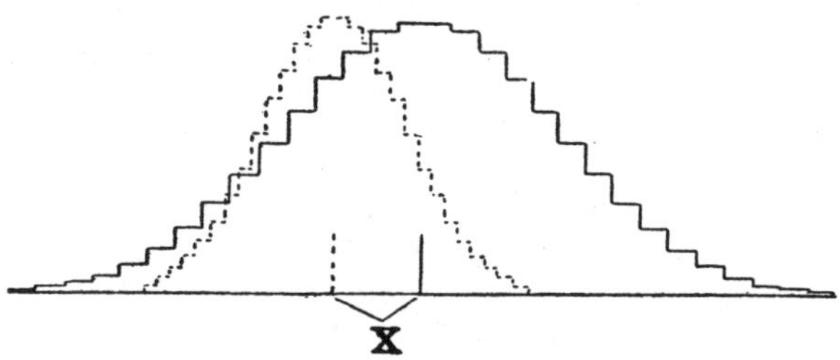

FIG. 86.

Given the fact that two groups are normally distributed and that the central tendency of the first is *X* plus the central tendency of the second, *X* being in terms of the variability of the first, what per cent. of the first group will exceed the central point for the second? The per cent. will equal 50 plus the per cent. included between the central point and a point *X* above it. (See Fig. 86.) This is, of course, given directly by the table. For instance, let group 1 have and $\sigma_{\text{dis.}} = 4$. Let group 2 have Av. = 8. The difference + 2 equals $.5\sigma$ (of distribution of group 1). The percentage of group 1 exceeding the average for group 2 will be 50 + 19.15 or 69.15 per cent.

When the first group is inferior to the second, the calculation is the same, replacing 50 per cent. plus by 50 per cent. minus.

68. If boys in spelling average 18.6 with $\sigma_{\text{dis.}} = 2.4$, and girls average 20.0, what per cent. of boys will reach or exceed the average for girls?

69. If the per cent. of attendance to enrollment in cities averages 74 with a $\text{P. E.}_{\text{dis.}}$ of 8.6, and the same trait in towns averages 64, what per cent. of cities will reach or exceed the average for towns?

70. If the median strength of 10-year-old boys is 16.2 with $\sigma_{\text{dis.}} = 2.1$, and the median strength of 11-year-old boys is 17.4, what per cent. of 10-year-olds will be stronger than the median 11-year-olds?

## CHAPTER XII.

### SOURCES OF ERROR IN MEASUREMENTS.

So far our supposition has been that the measures with which we start are accurate representatives of the fact measured, that *A* really did misspell the word which we score misspelled, that *B* did really take the .150 sec. to react which the chronoscope recorded, that the school enrollment and average attendance given for cities in the U. S. Commissioner's report give the real facts, that the number of children recorded in certain genealogy books for certain families were the real numbers. Our problem has been to make the best use of the data and introduce no error in manipulating them. But that a measure should thus perfectly represent a fact, the fact must be measured by a perfect instrument used by an infallible observer. In reality, any measure is a compound of a fact and the errors which the instrument and observer will surely make.

These errors may be *constant* or *variable.* A constant error is one tending more in one direction than the other. A watch that is too slow, a tendency of school superintendents to make the attendance record too high, are examples. Variable or chance errors are those tending in the long run to make the amount lower as often and as much as higher. The unevenness in action of a delicate balance due to dust, air currents, etc., the errors in addition made by the clerks in a superintendent's office, are examples.

Variable errors do not make any measure unfair, but only less exact and less reliable. If a body is weighed by an instrument which fluctuates so as to give 156.1, 156.2, 156.3, 156.3, 156.3, 156.3, 156.4, 156.4 and 156.4 in nine measurements, but is known not to weigh too light or heavy, 156.3 is a true measure, but the 156.3 only means between 156.25 and 156.35 and there is a slight chance of its being 156.2 or 156.4 (about 1 chance in 500).

If, on the contrary, a body is weighed by an instrument which fluctuates so little as to give 156.298, 156.299, 156.300, 156.300, 156.300, 156.301, 156.301 and 156.301, and which is known not to weigh too light or heavy, the 156.300 means between 156.2995 and 156.3005 and there is now certainty that the measure is not so low as 156.2 or so high as 156.4. Indeed, there is certainty that it is between 156.298 and 156.302.

There is no great advantage in decreasing the amount of the variable error by using more delicate instruments or more care in observing, unless the precision and reliability thereby obtained can be preserved in the further use of the measurements. The advantage that there is consists in the moral and intellectual training one gets and in the possibility that the measures may later be used for purposes other than one expects.

If we wish to get *A*'s average error in trying to equal a 100-mm. line, measurements may be made with the aid of a glass to $\frac{1}{10}$ mm., but the variation between *A*'s separate trials is so great that the larger error due to measuring each line so roughly as into $\frac{1}{2}$ mms. is insignificant. Indeed, measurements to a millimeter really do as well. If we wish to compare the reaction time of 1,000 boys with that of 1,000 girls, the median of 10 times being taken for each individual, measures in hundredths of seconds will do as well as measurements in thousandths.

Much time may be wasted in refining measurements in cases where no advantage accrues. And much ignorance is shown by the many students who disparage all measurements that are subject to a large variable error. They either do not know or forget that the reliability of a measure is due to the number of cases as well as to their variability, and that in the more complex and subtle mental traits it is always practicable to increase the number of measurements, but often impossible to make them less subject to variable errors. They also forget that the natural and real variability of the fact itself is often so large as to make the variability due to errors of instruments and observation practically negligible.

Constant errors, on the other hand, *are never negligible.*

The errors we make in interpreting handwriting would not, in a comparison of 1,000 boys with 1,000 girls in spelling ability, be worth spending a day on, even if thereby one could rectify them all, but if the teachers of the girls pronounced the words more clearly and phonetically than those of the boys, it would be necessary to discuss the proper discount or give up all hopes of precision. That a genealogist by mistake sometimes writes 4 or 7 matters practically *nil* to the student of vital statistics, but the genealogist's constant tendency to omit more children than he adds because of the difficulty of getting complete family records, is of the utmost importance.

Increasing the number of measures has here no beneficial influence. In certain cases increasing the number of observers may, namely, when the constant error of one observer is offset by the constant error *in the opposite direction* of another observer. If, that is, there is an error of prejudice or tendency constant for any one observer, but varying in direction by chance among a group of observers, what is a constant error for one becomes a variable error for a group, and is no longer a source of misleading, but only of lessened reliability. For instance, if any one person, even an expert judge, should rank 100 men in order for morality or efficiency or intellect, the results would probably have a constant error due to the undue weight he would put upon certain evidence; but if we took the median of the rankings given by ten or twelve expert judges, the error would in the main be only a chance error, for the prejudice of one would offset the prejudice of another.

The sources of constant errors in mental measurements are so numerous and so specialized for different kinds of facts that it is impossible to forearm the student against them here. Skill in avoiding them is due to capacity and watchfulness far more than to knowledge of any formal rules. It is, however, practically wise to test any result which may be affected by some constant error by using different methods of measurement, and to examine the means of selecting cases for measurement with the utmost care. The tendency to bias or to blunder is much more likely to make one select unfair cases than to make one measure them unfairly.

There is also a source of error which is perhaps in strictness an error in inference, but which from another point of view may be regarded as an error in measurement and so as relevant to the topics of this book. In measuring, say the spelling ability of a number of individuals whom we wish to compare, we assume that the achievement of each is a measure of the spelling ability of each. But *A* and *B* may have been seated where they did not hear the words pronounced so well as did *C* and *D*. *E* and *F* may have had headaches, while *G* and *H* were cheerful and bright. There exist errors due in the first example to outer physical conditions and in the second to inner or psychological conditions. To compare *A*, *B*, *C*, etc., in spelling ability, every extrinsic condition influencing that ability should be alike for all. Otherwise we are led into errors, which may be called errors of inferring an ability *in abstracto* from its manifestation under particular conditions, or of measuring a fact with a constant error of condition. It will be simpler to treat separately errors due to physical conditions and errors due to mental conditions.

Errors due to physical conditions can be prevented by making the conditions identical, or turned into relatively harmless variable errors by measuring each individual a number of times under conditions chosen at random. It would seem at first sight best to make conditions identical wherever practicable. This rule probably does hold for physical measurements, but there are certain disadvantages in this procedure in mental measurements. Too much artificiality and restraint in conditions often lead to an unusual and perturbed state of mind in the person measured, such that the thing one measures is likely to be a thing which would never occur in the ordinary course of the person's life. Measuring precisely a fact which you do not want is worse than measuring inexactly the fact you do want.

For instance, measurements of spelling under the unequal conditions of a schoolroom would, in spite of them, be better than measurements from 10-year-olds made to stand one at a time in the soundproof room of a laboratory with head exactly 50 centimeters from a phonograph which pronounced the words for them to spell. The last method would give identity of physical conditions, but would measure insensibility to strange surroundings and treatment and ability to attend to and interpret the phonograph's noises perhaps more than it would spelling ability.

Errors due to mental conditions can not be prevented with surety by making the conditions identical, for it is not in the power of the observer to control the mental conditions of the person measured. The best that can be done is to avoid any probable cause of difference in them and to take the subjects' reports as to what their mental conditions are. But mental conditions vary greatly even despite the apparent absence of causes for difference ; and the reports of mental condition from untrained self-observers must be vague, subject to constant errors and always from a personal standard of comparison incommensurate with that of any other individual. Though *A* says, 'I am tired,' and *B* says, 'I am not,' their feelings of fatigue may be equal. We do not take untrained individuals' opinions as facts elsewhere in science, and have no right to do so here. The more reliable procedure would be to eliminate the influence of the variability of inner conditions by random choice from among them rather than to pretend to eliminate the variation itself.

It is also a fair question whether the attempt to make all the mental conditions except the one to be measured alike in the persons to be compared, does not commonly result in so much unnaturalness of the sort against which protest was made a page back, as to do more harm than good. Attempted restriction of mental conditions surely disturbs any one even more than restriction of physical conditions.

Success in eliminating disturbing conditions is not attainable as a result of knowledge of any fixed rules, but only through a happy ingenuity in devising experiments, arranging observations and selecting data. We can, however, be careful, after securing the best measurements that we can, to distinguish sharply between the actual measurement of the fact under certain conditions, on the one hand, and on the other the inferences that we may be tempted to make about the fact in general or apart from those particular conditions. It is not undesirable to make inferences, but it is highly undesirable to confuse them with measurements or to leave them without critical scrutiny.

Much more might well be said with regard to the sources of error prevalent in studies of human nature, but the proper bounds of an introduction, not to the logic or general method of the mental sciences, but only to their statistical problems, have already been passed.

### *Weighting Results.*

Different sources of information concerning any one quantity may give it differing amounts, and these sources may be of unequal reliability. It is, then, desirable to allow more weight to the more trustworthy sources in deciding what amount is the most probable for the quantity. For instance, if an expert in physical anthropology measured *A*'s head and scored his cephalic index .81, while an ordinary person scored it .80, we should choose the .81 rather than the .80, and, if we allowed something for each judgment, would perhaps take 80.8 as the figure, counting the anthropologist's result four times.

No care in weighting sources will do so much service as the

elimination of constant errors; and ideally no source with a constant error unallowed for should have any place in determining a result. Any source may deserve weight because of either numerical or qualitative strength. Its numerical strength is as the square root of the number of cases whose study it represents. Weighting for quality is bound in practice to be largely arbitrary, but this is not a great misfortune, for the result will rarely be altered appreciably by such differences in the system of weighting as reasonably competent students would make. For instance, *A*, *B* and *C* with the same general problem use different methods and get as a certain correlation coefficient .60, .50 and .48 respectively. Suppose that we weight these sources 1, 1, and 1; 4, 4 and 5; 3, 4 and 5; and finally 4, 3 and 5. We have then, as the probable true coefficient, .5267, .5231, .5167 or .5250. Bowley gives a rule that is satisfactory for most cases that occur in practice, namely, to give your attention to eliminating constant errors and not to manipulating weights.* If results are weighted it is always well to give them in their unweighted form as well and leave the opportunity open for any critic to weight them as he judges proper.

* 'In calculating averages give all your care to making the items free from bias and leave the weights to take care of themselves.' 'Elements of Statistics,' p. 118.

# CHAPTER XIII.

## CONCLUSION. REFERENCES FOR FURTHER STUDY.

I TRUST that the reader has been impressed by now with the fact that the theory of mental measurements is no display of mathematical pedantry or subtle juggling with figures, but on the contrary is simple common sense. The chief lessons of this book are in fact simple applications of the most elementary logic. They may be summed up in the form of warnings against certain fallacies common in the quantitative treatment of mental facts, *viz.:*

1. Accepting guessed equality or mere verbal likeness in place of real equality.

2. Using quantities on a scale without consideration of the meaning of the scale's zero point.

3. Dealing carelessly with totals the constitution of which is unknown.

4. Using an average to represent a series of individual measures regardless of their distribution.

5. Estimating a total series from individual measures numerically insufficient or so selected as to actually misrepresent it.

6. Estimating differences by ambiguous measures.

7. Using a difference between or change in averages to represent a series of individual differences or changes. (7 is essentially the same fallacy as 4.)

If the reader has been rendered immune to these errors, has acquired facility and confidence in the manipulation of measurements, and has learned to discard guess work and crude arithmetic in favor of accurate and modern methods of measuring facts and relationships, the purpose of this book has been amply fulfilled.

It is desirable that the student who has been thus introduced to statistical methods should proceed to study samples of their concrete application to problems in the mental sciences and, in case he has the necessary mathematical interest and training, that he should study the abstract properties of different types of distribution, the derivation of statistical formulæ, the mathematical theory of correlation and other topics in pure statistics. To these ends the references given below may be useful.

These will be grouped in accordance with the different interests which may be supposed to dominate the quantitative studies of readers of this introduction, under psychology, education, economics and social science, anthropometry, vital statistics and biology. A few references to the most easily understood articles on pure statistics will form a group by themselves. The order in which the references for each topic are given is that in which the student may profitably read them.

*Psychology.*

'On the Perception of Small Differences.' By G. S. Fullerton and and J. McK. Cattell. No. 2 of the Philosophical Series of the Publications of the University of Pennsylvania, May, 1892. The University of Pennsylvania Press, Philadelphia.

Quantitative exactitude was first sought by psychologists in the case of the ability to perceive differences. The monograph by Fullerton and Cattell gives a clear account of the common methods of estimating quantitatively psycho-physical relationships, *viz.*, the method of the just noticeable difference, the method of right and wrong cases, the method of average error and the method of mean gradation. It also represents an investigation made with full consciousness and appreciation of the special problems of variable phenomena. It is thus the best introduction to the special problems in mental measurement which confront the student of psycho-physics.

TABLE FOR DETERMINING THE PROBABLE ERROR FROM THE PERCENTAGE OF RIGHT CASES AND AMOUNT OF DIFFERENCE.*

| % r. | $\frac{\Delta}{P. E.}$ | % r. | $\frac{\Delta}{P. E.}$ | % r. | $\frac{\Delta}{P. E.}$ | % r. | $\frac{\Delta}{P. E.}$ | % r. | $\frac{\Delta}{P. E.}$ |
|---|---|---|---|---|---|---|---|---|---|
| 50 | .00 | 60 | .38 | 70 | .78 | 80 | 1.25 | 90 | 1.90 |
| 51 | .04 | 61 | .41 | 71 | .82 | 81 | 1.30 | 91 | 1.99 |
| 52 | .07 | 62 | .45 | 72 | .86 | 82 | 1.36 | 92 | 2.08 |
| 53 | .11 | 63 | .49 | 73 | .91 | 83 | 1.41 | 93 | 2.19 |
| 54 | .15 | 64 | .53 | 74 | .95 | 84 | 1.47 | 94 | 2.31 |
| 55 | .19 | 65 | .57 | 75 | 1.00 | 85 | 1.54 | 95 | 2.44 |
| 56 | .22 | 66 | .61 | 76 | 1.05 | 86 | 1.60 | 96 | 2.60 |
| 57 | .26 | 67 | .65 | 77 | 1.10 | 87 | 1.67 | 97 | 2.79 |
| 58 | .30 | 68 | .69 | 78 | 1.14 | 88 | 1.74 | 98 | 3.05 |
| 59 | .34 | 69 | .74 | 79 | 1.20 | 89 | 1.82 | 99 | 3.45 |

*From page 16 of 'The Perception of Small Differences,' by Fullerton and Cattell.

The table for estimating the P. E. from the percentage of right cases for a given difference is so frequently useful that I reprint it here for the sake of those to whom the monograph may be inaccessible.

'Hereditary Genius.' By Francis Galton. Chapters 1, 2 and 3.

'The Correlation of Mental and Physical Measurements.' By Clark Wissler. Monograph Supplement, No. 16, to the *Psychological Review*.

'Natural Inheritance.' By Francis Galton. Chapters 8 and 9.

'Statistics of American Psychologists.' By J. McKeen Cattell. *American Journal of Psychology*, Vol. XIV., pp. 310–328.

The last two studies illustrate the importance of measures by relative position. Since such measures are likely to be of great service in the social sciences and in scientific studies of history and literature, these articles may well be examined by other than psychological students.

*Education.*

'The Age of Graduation from College.' By Winfield Scott Thomas. *Popular Science Monthly*, June, 1903.

The article by Thomas, though extremely simple, is a most useful illustration of the value of other measures than the average for a central tendency and of the significance of measures of variability.

'The Correlations of the Abilities Involved in Secondary School Work.' By W. P. Burris. In Heredity, Correlation and Sex Differences in School Abilities; *Columbia Contributions to Philosophy, Psychology and Education*, Vol. XI., No. 2.

This article represents a condensed report. Hence the method used is incompletely described and the original data are omitted. The article is, however, valuable as a suggestion of the susceptibility of even complex educational problems to exact quantitative study. In spite of the wealth of material at hand in school reports, teacher's records and the like, the author can find no better samples of the use of modern statistical methods in educational science than these two slight studies.

*Economics and Social Science.*

'Elements of Statistics.' By A. L. Bowley.

This book, besides giving a general account of statistical procedure in economics, contains many samples of facts and relations adequately described and a comparatively simple account of the application of the theory of probability to measurements of facts.

'Notes on the History of Pauperism in England and Wales from 1850, treated by the method of frequency-curves; with an introduction on the method.' By G. Udney Yule, *Journal of the Royal Statistical Society*, June, 1896.

'On the Correlation of Total Pauperism with Proportion of Outdoor Relief.' By G. Udney Yule. *Economic Journal*, December, 1895, and December, 1896.

'An Investigation into the Causes of Changes in Pauperism, in England Chiefly during the last Two Intercensal Periods.' By G. Udney Yule. *Journal of the Royal Statistical Society*, June, 1899.

Professor Yule's articles on pauperism represent the application of modern methods of measurement to the economic and social sciences. They illustrate the advantages to be gained in these sciences from dealing with total distributions rather than averages and from using appropriate methods of measuring variable relationships. By means of Pearson coefficients of correlation, Professor Yule was able to turn certain data on pauperism to a new use. Care in the mathematical handling of the measures used is also well shown. In respect to wise choice of units and a vivid sense of the concrete facts represented by the measures, the articles are more questionable.

### *Anthropometry.*

'Natural Inheritance.' By Francis Galton. Chapters 1–7.

'The Growth of United States Naval Cadets.' By H. G. Beyer. *Proceedings of the United States Naval Institute.* Vol. 21 (1895), pp. 297–333.

The present activity on the part of English men of science in developing methods of exact measurement of variable phenomena had its source in Galton's work. This book is therefore a fitting introduction for the student because of its historical importance as well as the relative simplicity of its mathematics. Dr. Beyer's article is still simpler in its manner of presentation, but is unfortunately inaccessible to most students.

'The Growth of Boys.' By C. Wissler. *American Anthropologist*, New Series, Vol. V., No. 1.

The article by Wissler reports one of the very few studies of change in which changes themselves are measured. It demonstrates in a most elegant manner the law of compensation by which relatively slow growth up to a certain age implies relatively rapid growth thereafter. If the material had been lumped into undistributed averages in the customary way none of the author's conclusions could have been reached.

'The Cephalic Index.' By Franz Boas. *American Anthropologist*, New Series, Vol. I., pp. 448–461.

'The Growth of Toronto Children.' By Franz Boas. Report of the United States Commissioner of Education for 1896–97, Vol. 2, pp. 1541–1599.

'On the Variability and Correlation of the Hand.' By M. A. Whiteley and Karl Pearson. *Proceedings of the Royal Society of London*, Vol. 65, pp. 126–151.

'On the Variability and Correlation of the Hand.' By M. A. Lewenz and M. A. Whiteley. *Biometrika*, Vol. I.

The first article by Boas is especially interesting as an illustration of the uses of exact statistical methods in elucidating causes. The second article by Boas and the articles on the anatomy of the hand, report studies made with extreme quantitative refinement and presented in full detail.

*Vital Statistics.*

'The Chances of Death.' By Karl Pearson. In a volume with the same title.

'Zur Theorie der Massenerscheinungen in der Menschlichen Gesellschaft.' By W. Lexis.

'On the Inheritance of the Duration of Life.' By Mary Beeton and Karl Pearson. *Biometrika*, Vol. I.

*Biology.*

'Statistical Methods.' By C. B. Davenport.

'Die Methode der Variations-Statistik.' G. Duncker. *Arch. f. Entwickelungs-Mechan. d. Organismen*, VIII., 112–183.

For further references see the bibliographies given by Davenport and Duncker.

*Pure Statistics.*

'The Principles of Science.' W. S. Jevons.

'The Logic of Chance.' J. Venn.

The chapters on permutations, combinations and probability in any standard algebra.

'History of the Theory of Probability.' I. Todhunter.

'The Method of Least Squares.' M. Merriman.

'Hereditary Genius.' F. Galton. Chapters 1–3.

'Natural Inheritance.' F. Galton. Chapters 1–7.

'Lettres sur la Probabilité.' A. Quetelet. (Difficult of access.)

'Elements of Statistics.' A. L. Bowley. Part II.

'Grammar of Science' (second edition). Karl Pearson. Chapters X.–XI.

'Theorie der Bevolkerungs und Moralstatistik.' W. Lexis. Chapter VI.

'On the Theory of Correlation.' G. U. Yule. *Journal of the Royal Statistical Society*, Vol. 60, pp. 812–854.

'Collektivmasslehre.' G. T. Fechner.

'The Proof and Measurement of Association Between Two Things.' C. Spearman. *American Journal of Psychology*, January, 1904, Vol. XV., pp. 72–101.

Material for more advanced study of pure statistics will be found in the writings of Franz Boas, H. Bruns, F. Y. Edgeworth, W. Lexis, G. Lipps, Karl Pearson, W. F. Sheppard, H. Westergaard and G. U. Yule.

The contributions of the English students of pure statistics will be found chiefly in the *Philosophical Transactions of the Royal Society of London*, in the *Proceedings* of the same society, in the *Journal of the Royal Statistical Society*, in *Biometrika*, and in the *London, Edinburgh, and Dublin Philosophical Magazine and Journal of Science.*

Special lists of references to both pure and applied statistics will be found in Bowley's 'Elements of Statistics,' Davenport's 'Statistical Methods' and Duncker's 'Methode der Variations-Statistik.'

# APPENDIX I.

## A MULTIPLICATION TABLE UP TO 100 × 100.

THE reader's attention has already been called to Crelle's *Rechentafeln*, a multiplication table up to 1000 × 1000. It saves much time, replaces mental work by finger and eye work, and decreases errors in calculation. Crelle's table, however, makes a book some 9 by 14 inches, weighing several pounds. The table that follows is a modification of Crelle's table, but runs only to 100 × 100. For work with these smaller numbers and for approximate calculations, it is more rapid than the longer table and is so arranged as to be easier for the eyes.

Its uses will be apparent upon examination, but the reader should note that it serves for division as well as for multiplication. In dividing, one of course finds the divisor in the row of figures in heavy faced type at the top of the page, hunts for the dividend in the column beneath it, and, this being found, obtains the quotient in the figure in heavy-faced type at the side of the page. Thus to divide 684 by 38, one looks under 38, finds 684 and opposite it, at the side of the page, 18, the answer. Again to divide 1,600 by 38, one looks under 38, finds 1596 to be the nearest number, and so the nearest two-figure answer to be 42. If one needed greater precision, he could divide the remainder 4.0 by 38, getting 0.1, and then the remainder .2000, getting .0052, or 42.1052, and so on to any desired precision.

| | 2 | 3 | 4 | 5 | 6 | 7 | 8 | 9 | 10 | |
|---|---|---|---|---|---|---|---|---|---|---|
| 1 | 2 | 3 | 4 | 5 | 6 | 7 | 8 | 9 | 10 | 1 |
| 2 | 4 | 6 | 8 | 10 | 12 | 14 | 16 | 18 | 20 | 2 |
| 3 | 6 | 9 | 12 | 15 | 18 | 21 | 24 | 27 | 30 | 3 |
| 4 | 8 | 12 | 16 | 20 | 24 | 28 | 32 | 36 | 40 | 4 |
| 5 | 10 | 15 | 20 | 25 | 30 | 35 | 40 | 45 | 50 | 5 |
| 6 | 12 | 18 | 24 | 30 | 36 | 42 | 48 | 54 | 60 | 6 |
| 7 | 14 | 21 | 28 | 35 | 42 | 49 | 56 | 63 | 70 | 7 |
| 8 | 16 | 24 | 32 | 40 | 48 | 56 | 64 | 72 | 80 | 8 |
| 9 | 18 | 27 | 36 | 45 | 54 | 63 | 72 | 81 | 90 | 9 |
| 10 | 20 | 30 | 40 | 50 | 60 | 70 | 80 | 90 | 100 | 10 |
| 11 | 22 | 33 | 44 | 55 | 66 | 77 | 88 | 99 | 110 | 11 |
| 12 | 24 | 36 | 48 | 60 | 72 | 84 | 96 | 108 | 120 | 12 |
| 13 | 26 | 39 | 52 | 65 | 78 | 91 | 104 | 117 | 130 | 13 |
| 14 | 28 | 42 | 56 | 70 | 84 | 98 | 112 | 126 | 140 | 14 |
| 15 | 30 | 45 | 60 | 75 | 90 | 105 | 120 | 135 | 150 | 15 |
| 16 | 32 | 48 | 64 | 80 | 96 | 112 | 128 | 144 | 160 | 16 |
| 17 | 34 | 51 | 68 | 85 | 102 | 119 | 136 | 153 | 170 | 17 |
| 18 | 36 | 54 | 72 | 90 | 108 | 126 | 144 | 162 | 180 | 18 |
| 19 | 38 | 57 | 76 | 95 | 114 | 133 | 152 | 171 | 190 | 19 |
| 20 | 40 | 60 | 80 | 100 | 120 | 140 | 160 | 180 | 200 | 20 |
| 21 | 42 | 63 | 84 | 105 | 126 | 147 | 168 | 189 | 210 | 21 |
| 22 | 44 | 66 | 88 | 110 | 132 | 154 | 176 | 198 | 220 | 22 |
| 23 | 46 | 69 | 92 | 115 | 138 | 161 | 184 | 207 | 230 | 23 |
| 24 | 48 | 72 | 96 | 120 | 144 | 168 | 192 | 216 | 240 | 24 |
| 25 | 50 | 75 | 100 | 125 | 150 | 175 | 200 | 225 | 250 | 25 |
| 26 | 52 | 78 | 104 | 130 | 156 | 182 | 208 | 234 | 260 | 26 |
| 27 | 54 | 81 | 108 | 135 | 162 | 189 | 216 | 243 | 270 | 27 |
| 28 | 56 | 84 | 112 | 140 | 168 | 196 | 224 | 252 | 280 | 28 |
| 29 | 58 | 87 | 116 | 145 | 174 | 203 | 232 | 261 | 290 | 29 |
| 30 | 60 | 90 | 120 | 150 | 180 | 210 | 240 | 270 | 300 | 30 |
| 31 | 62 | 93 | 124 | 155 | 186 | 217 | 248 | 279 | 310 | 31 |
| 32 | 64 | 96 | 128 | 160 | 192 | 224 | 256 | 288 | 320 | 32 |
| 33 | 66 | 99 | 132 | 165 | 198 | 231 | 264 | 297 | 330 | 33 |
| 34 | 68 | 102 | 136 | 170 | 204 | 238 | 272 | 306 | 340 | 34 |
| 35 | 70 | 105 | 140 | 175 | 210 | 245 | 280 | 315 | 350 | 35 |
| 36 | 72 | 108 | 144 | 180 | 216 | 252 | 288 | 324 | 360 | 36 |
| 37 | 74 | 111 | 148 | 185 | 222 | 259 | 296 | 333 | 370 | 37 |
| 38 | 76 | 114 | 152 | 190 | 228 | 266 | 304 | 342 | 380 | 38 |
| 39 | 78 | 117 | 156 | 195 | 234 | 273 | 312 | 351 | 390 | 39 |
| 40 | 80 | 120 | 160 | 200 | 240 | 280 | 320 | 360 | 400 | 40 |
| 41 | 82 | 123 | 164 | 205 | 246 | 287 | 328 | 369 | 410 | 41 |
| 42 | 84 | 126 | 168 | 210 | 252 | 294 | 336 | 378 | 420 | 42 |
| 43 | 86 | 129 | 172 | 215 | 258 | 301 | 344 | 387 | 430 | 43 |
| 44 | 88 | 132 | 176 | 220 | 264 | 308 | 352 | 396 | 440 | 44 |
| 45 | 90 | 135 | 180 | 225 | 270 | 315 | 360 | 405 | 450 | 45 |
| 46 | 92 | 138 | 184 | 230 | 276 | 322 | 368 | 414 | 460 | 46 |
| 47 | 94 | 141 | 188 | 235 | 282 | 319 | 376 | 423 | 470 | 47 |
| 48 | 96 | 144 | 192 | 240 | 288 | 336 | 384 | 432 | 480 | 48 |
| 49 | 98 | 147 | 196 | 245 | 294 | 343 | 392 | 441 | 490 | 49 |
| 50 | 100 | 150 | 200 | 250 | 300 | 350 | 400 | 450 | 500 | 50 |
| | 2 | 3 | 4 | 5 | 6 | 7 | 8 | 9 | 10 | |

| | 2 | 3 | 4 | 5 | 6 | 7 | 8 | 9 | 10 | |
|---|---|---|---|---|---|---|---|---|---|---|
| **51** | 102 | 153 | 204 | 255 | 306 | 357 | 408 | 459 | 510 | **51** |
| **52** | 104 | 156 | 208 | 260 | 312 | 364 | 416 | 468 | 520 | **52** |
| **53** | 106 | 159 | 212 | 265 | 318 | 371 | 424 | 477 | 530 | **53** |
| **54** | 108 | 162 | 216 | 270 | 324 | 378 | 432 | 486 | 540 | **54** |
| **55** | 110 | 165 | 220 | 275 | 330 | 385 | 440 | 495 | 550 | **55** |
| **56** | 112 | 168 | 224 | 280 | 336 | 392 | 448 | 504 | 560 | **56** |
| **57** | 114 | 171 | 228 | 285 | 342 | 399 | 456 | 513 | 570 | **57** |
| **58** | 116 | 174 | 232 | 290 | 348 | 406 | 464 | 522 | 580 | **58** |
| **59** | 118 | 177 | 236 | 295 | 354 | 413 | 472 | 531 | 590 | **59** |
| **60** | 120 | 180 | 240 | 300 | 360 | 420 | 480 | 540 | 600 | **60** |
| **61** | 122 | 183 | 244 | 305 | 366 | 427 | 488 | 549 | 610 | **61** |
| **62** | 124 | 186 | 248 | 310 | 372 | 434 | 496 | 558 | 620 | **62** |
| **63** | 126 | 189 | 252 | 315 | 378 | 441 | 504 | 567 | 630 | **63** |
| **64** | 128 | 192 | 256 | 320 | 384 | 448 | 512 | 576 | 640 | **64** |
| **65** | 130 | 195 | 260 | 325 | 390 | 455 | 520 | 585 | 650 | **65** |
| **66** | 132 | 198 | 264 | 330 | 396 | 462 | 528 | 594 | 660 | **66** |
| **67** | 134 | 201 | 268 | 335 | 402 | 469 | 536 | 603 | 670 | **67** |
| **68** | 136 | 204 | 272 | 340 | 408 | 476 | 544 | 612 | 680 | **68** |
| **69** | 138 | 207 | 276 | 345 | 414 | 483 | 552 | 621 | 690 | **69** |
| **70** | 140 | 210 | 280 | 350 | 420 | 490 | 560 | 630 | 700 | **70** |
| **71** | 142 | 213 | 284 | 355 | 426 | 497 | 568 | 639 | 710 | **71** |
| **72** | 144 | 216 | 288 | 360 | 432 | 504 | 576 | 648 | 720 | **72** |
| **73** | 146 | 219 | 292 | 365 | 438 | 511 | 584 | 657 | 730 | **73** |
| **74** | 148 | 222 | 296 | 370 | 444 | 518 | 592 | 666 | 740 | **74** |
| **75** | 150 | 225 | 300 | 375 | 450 | 525 | 600 | 675 | 750 | **75** |
| **76** | 152 | 228 | 304 | 380 | 456 | 532 | 608 | 684 | 760 | **76** |
| **77** | 154 | 231 | 308 | 385 | 462 | 539 | 616 | 693 | 770 | **77** |
| **78** | 156 | 234 | 312 | 390 | 468 | 546 | 624 | 702 | 780 | **78** |
| **79** | 158 | 237 | 316 | 395 | 474 | 553 | 632 | 711 | 790 | **79** |
| **80** | 160 | 240 | 320 | 400 | 480 | 560 | 640 | 720 | 800 | **80** |
| **81** | 162 | 243 | 324 | 405 | 486 | 567 | 648 | 729 | 810 | **81** |
| **82** | 164 | 246 | 328 | 410 | 492 | 574 | 656 | 738 | 820 | **82** |
| **83** | 166 | 249 | 332 | 415 | 498 | 581 | 664 | 747 | 830 | **83** |
| **84** | 168 | 252 | 336 | 420 | 504 | 588 | 672 | 756 | 840 | **84** |
| **85** | 170 | 255 | 340 | 425 | 510 | 595 | 680 | 765 | 850 | **85** |
| **86** | 172 | 258 | 344 | 430 | 516 | 602 | 688 | 774 | 860 | **86** |
| **87** | 174 | 261 | 348 | 435 | 522 | 609 | 696 | 783 | 870 | **87** |
| **88** | 176 | 264 | 352 | 440 | 528 | 616 | 704 | 792 | 880 | **88** |
| **89** | 178 | 267 | 356 | 445 | 534 | 623 | 712 | 801 | 890 | **89** |
| **90** | 180 | 270 | 360 | 450 | 540 | 630 | 720 | 810 | 900 | **90** |
| **91** | 182 | 273 | 364 | 455 | 546 | 637 | 728 | 819 | 910 | **91** |
| **92** | 184 | 276 | 368 | 460 | 552 | 644 | 736 | 828 | 920 | **92** |
| **93** | 186 | 279 | 372 | 465 | 558 | 651 | 744 | 837 | 930 | **93** |
| **94** | 188 | 282 | 376 | 470 | 564 | 658 | 752 | 846 | 940 | **94** |
| **95** | 190 | 285 | 380 | 475 | 570 | 665 | 760 | 855 | 950 | **95** |
| **96** | 192 | 288 | 384 | 480 | 576 | 672 | 768 | 864 | 960 | **96** |
| **97** | 194 | 291 | 388 | 485 | 582 | 679 | 776 | 873 | 970 | **97** |
| **98** | 196 | 294 | 392 | 490 | 588 | 686 | 784 | 882 | 980 | **98** |
| **99** | 198 | 297 | 396 | 495 | 594 | 693 | 792 | 891 | 990 | **99** |
| **100** | 200 | 300 | 400 | 500 | 600 | 700 | 800 | 900 | 1000 | **100** |
| | **2** | **3** | **4** | **5** | **6** | **7** | **8** | **9** | **10** | |

| | 11 | 12 | 13 | 14 | 15 | 16 | 17 | 18 | 19 | 20 | |
|---|---|---|---|---|---|---|---|---|---|---|---|
| 1 | 11 | 12 | 13 | 14 | 15 | 16 | 17 | 18 | 19 | 20 | 1 |
| 2 | 22 | 24 | 26 | 28 | 30 | 32 | 34 | 36 | 38 | 40 | 2 |
| 3 | 33 | 36 | 39 | 42 | 45 | 48 | 51 | 54 | 57 | 60 | 3 |
| 4 | 44 | 48 | 52 | 56 | 60 | 64 | 68 | 72 | 76 | 80 | 4 |
| 5 | 55 | 60 | 65 | 70 | 75 | 80 | 85 | 90 | 95 | 100 | 5 |
| 6 | 66 | 72 | 78 | 84 | 90 | 96 | 102 | 108 | 114 | 120 | 6 |
| 7 | 77 | 84 | 91 | 98 | 105 | 112 | 119 | 126 | 133 | 140 | 7 |
| 8 | 88 | 96 | 104 | 112 | 120 | 128 | 136 | 144 | 152 | 160 | 8 |
| 9 | 99 | 108 | 117 | 126 | 135 | 144 | 153 | 162 | 171 | 180 | 9 |
| 10 | 110 | 120 | 130 | 140 | 150 | 160 | 170 | 180 | 190 | 200 | 10 |
| 11 | 121 | 132 | 143 | 154 | 165 | 176 | 187 | 198 | 209 | 220 | 11 |
| 12 | 132 | 144 | 156 | 168 | 180 | 192 | 204 | 216 | 228 | 240 | 12 |
| 13 | 143 | 156 | 169 | 182 | 195 | 208 | 221 | 234 | 247 | 260 | 13 |
| 14 | 154 | 168 | 182 | 196 | 210 | 224 | 238 | 252 | 266 | 280 | 14 |
| 15 | 165 | 180 | 195 | 210 | 225 | 240 | 255 | 270 | 285 | 300 | 15 |
| 16 | 176 | 192 | 208 | 224 | 240 | 256 | 272 | 288 | 304 | 320 | 16 |
| 17 | 187 | 204 | 221 | 238 | 255 | 272 | 289 | 306 | 323 | 340 | 17 |
| 18 | 198 | 216 | 234 | 252 | 270 | 288 | 306 | 324 | 342 | 360 | 18 |
| 19 | 209 | 228 | 247 | 266 | 285 | 304 | 323 | 342 | 361 | 380 | 19 |
| 20 | 220 | 240 | 260 | 280 | 300 | 320 | 340 | 360 | 380 | 400 | 20 |
| 21 | 231 | 252 | 273 | 294 | 315 | 336 | 357 | 378 | 399 | 420 | 21 |
| 22 | 242 | 264 | 286 | 308 | 330 | 352 | 374 | 396 | 418 | 440 | 22 |
| 23 | 253 | 276 | 299 | 322 | 345 | 368 | 391 | 414 | 437 | 460 | 23 |
| 24 | 264 | 288 | 312 | 336 | 360 | 384 | 408 | 432 | 456 | 480 | 24 |
| 25 | 275 | 300 | 325 | 350 | 375 | 400 | 425 | 450 | 475 | 500 | 25 |
| 26 | 286 | 312 | 338 | 364 | 390 | 416 | 442 | 468 | 494 | 520 | 26 |
| 27 | 297 | 324 | 351 | 378 | 405 | 432 | 459 | 486 | 513 | 540 | 27 |
| 28 | 308 | 336 | 364 | 392 | 420 | 448 | 476 | 504 | 532 | 560 | 28 |
| 29 | 319 | 348 | 377 | 406 | 435 | 464 | 493 | 522 | 551 | 580 | 29 |
| 30 | 330 | 360 | 390 | 420 | 450 | 480 | 510 | 540 | 570 | 600 | 30 |
| 31 | 341 | 372 | 403 | 434 | 465 | 496 | 527 | 558 | 589 | 620 | 31 |
| 32 | 352 | 384 | 416 | 448 | 480 | 512 | 544 | 576 | 608 | 640 | 32 |
| 33 | 363 | 396 | 429 | 462 | 495 | 528 | 561 | 594 | 627 | 660 | 33 |
| 34 | 374 | 408 | 442 | 476 | 510 | 544 | 578 | 612 | 646 | 680 | 34 |
| 35 | 385 | 420 | 455 | 490 | 525 | 560 | 595 | 630 | 665 | 700 | 35 |
| 36 | 396 | 432 | 468 | 504 | 540 | 576 | 612 | 648 | 684 | 720 | 36 |
| 37 | 407 | 444 | 481 | 518 | 555 | 592 | 629 | 666 | 703 | 740 | 37 |
| 38 | 418 | 456 | 494 | 532 | 570 | 608 | 646 | 684 | 722 | 760 | 38 |
| 39 | 429 | 468 | 507 | 546 | 585 | 624 | 663 | 702 | 741 | 780 | 39 |
| 40 | 440 | 480 | 520 | 560 | 600 | 640 | 680 | 720 | 760 | 800 | 40 |
| 41 | 451 | 492 | 533 | 574 | 615 | 656 | 697 | 738 | 779 | 820 | 41 |
| 42 | 462 | 504 | 546 | 588 | 630 | 672 | 714 | 756 | 798 | 840 | 42 |
| 43 | 473 | 516 | 559 | 602 | 645 | 688 | 731 | 774 | 817 | 860 | 43 |
| 44 | 484 | 528 | 572 | 616 | 660 | 704 | 748 | 792 | 836 | 880 | 44 |
| 45 | 495 | 540 | 585 | 630 | 675 | 720 | 765 | 810 | 855 | 900 | 45 |
| 46 | 506 | 552 | 598 | 644 | 690 | 736 | 782 | 828 | 874 | 920 | 46 |
| 47 | 517 | 564 | 611 | 658 | 705 | 752 | 799 | 846 | 893 | 940 | 47 |
| 48 | 528 | 576 | 624 | 672 | 720 | 768 | 816 | 864 | 912 | 960 | 48 |
| 49 | 539 | 588 | 637 | 686 | 735 | 784 | 833 | 882 | 931 | 980 | 49 |
| 50 | 550 | 600 | 650 | 700 | 750 | 800 | 850 | 900 | 950 | 1000 | 50 |
| | 11 | 12 | 13 | 14 | 15 | 16 | 17 | 18 | 19 | 20 | |

| | 11 | 12 | 13 | 14 | 15 | 16 | 17 | 18 | 19 | 20 | |
|---|---|---|---|---|---|---|---|---|---|---|---|
| **51** | 561 | 612 | 663 | 714 | 765 | 816 | 867 | 918 | 969 | 1020 | **51** |
| **52** | 572 | 624 | 676 | 728 | 780 | 832 | 884 | 936 | 988 | 1040 | **52** |
| **53** | 583 | 636 | 689 | 742 | 795 | 848 | 901 | 954 | 1007 | 1060 | **53** |
| **54** | 594 | 648 | 702 | 756 | 810 | 864 | 918 | 972 | 1026 | 1080 | **54** |
| **55** | 605 | 660 | 715 | 770 | 825 | 880 | 935 | 990 | 1045 | 1100 | **55** |
| **56** | 616 | 672 | 728 | 784 | 840 | 896 | 952 | 1008 | 1064 | 1120 | **56** |
| **57** | 627 | 684 | 741 | 798 | 855 | 912 | 969 | 1026 | 1083 | 1140 | **57** |
| **58** | 638 | 696 | 754 | 812 | 870 | 928 | 986 | 1044 | 1102 | 1160 | **58** |
| **59** | 649 | 708 | 767 | 826 | 885 | 944 | 1003 | 1062 | 1121 | 1180 | **59** |
| **60** | 660 | 720 | 780 | 840 | 900 | 960 | 1020 | 1080 | 1140 | 1200 | **60** |
| **61** | 671 | 732 | 793 | 854 | 915 | 976 | 1037 | 1098 | 1159 | 1220 | **61** |
| **62** | 682 | 744 | 806 | 868 | 930 | 992 | 1054 | 1116 | 1178 | 1240 | **62** |
| **63** | 693 | 756 | 819 | 882 | 945 | 1008 | 1071 | 1134 | 1197 | 1260 | **63** |
| **64** | 704 | 768 | 832 | 896 | 960 | 1024 | 1088 | 1152 | 1216 | 1280 | **64** |
| **65** | 715 | 780 | 845 | 910 | 975 | 1040 | 1105 | 1170 | 1235 | 1300 | **65** |
| **66** | 726 | 792 | 858 | 924 | 990 | 1056 | 1122 | 1188 | 1254 | 1320 | **66** |
| **67** | 737 | 804 | 871 | 938 | 1005 | 1072 | 1139 | 1206 | 1273 | 1340 | **67** |
| **68** | 748 | 816 | 884 | 952 | 1020 | 1088 | 1156 | 1224 | 1292 | 1360 | **68** |
| **69** | 759 | 828 | 897 | 966 | 1035 | 1104 | 1173 | 1242 | 1311 | 1380 | **69** |
| **70** | 770 | 840 | 910 | 980 | 1050 | 1120 | 1190 | 1260 | 1330 | 1400 | **70** |
| **71** | 781 | 852 | 923 | 994 | 1065 | 1136 | 1207 | 1278 | 1349 | 1420 | **71** |
| **72** | 792 | 864 | 936 | 1008 | 1080 | 1152 | 1224 | 1296 | 1368 | 1440 | **72** |
| **73** | 803 | 876 | 949 | 1022 | 1095 | 1168 | 1241 | 1314 | 1387 | 1460 | **73** |
| **74** | 814 | 888 | 962 | 1036 | 1110 | 1184 | 1258 | 1332 | 1406 | 1480 | **74** |
| **75** | 825 | 900 | 975 | 1050 | 1125 | 1200 | 1275 | 1350 | 1425 | 1500 | **75** |
| **76** | 836 | 912 | 988 | 1064 | 1140 | 1216 | 1292 | 1368 | 1444 | 1520 | **76** |
| **77** | 847 | 924 | 1001 | 1078 | 1155 | 1232 | 1309 | 1386 | 1463 | 1540 | **77** |
| **78** | 858 | 936 | 1014 | 1092 | 1170 | 1248 | 1326 | 1404 | 1482 | 1560 | **78** |
| **79** | 869 | 948 | 1027 | 1106 | 1185 | 1264 | 1343 | 1422 | 1501 | 1580 | **79** |
| **80** | 880 | 960 | 1040 | 1120 | 1200 | 1280 | 1360 | 1440 | 1520 | 1600 | **80** |
| **81** | 891 | 972 | 1053 | 1134 | 1215 | 1296 | 1377 | 1458 | 1539 | 1620 | **81** |
| **82** | 902 | 984 | 1066 | 1148 | 1230 | 1312 | 1394 | 1476 | 1558 | 1640 | **82** |
| **83** | 913 | 996 | 1079 | 1162 | 1245 | 1328 | 1411 | 1494 | 1577 | 1660 | **83** |
| **84** | 924 | 1008 | 1092 | 1176 | 1260 | 1344 | 1428 | 1512 | 1596 | 1680 | **84** |
| **85** | 935 | 1020 | 1105 | 1190 | 1275 | 1360 | 1445 | 1530 | 1615 | 1700 | **85** |
| **86** | 946 | 1032 | 1118 | 1204 | 1290 | 1376 | 1462 | 1548 | 1634 | 1720 | **86** |
| **87** | 957 | 1044 | 1131 | 1218 | 1305 | 1392 | 1479 | 1566 | 1653 | 1740 | **87** |
| **88** | 968 | 1056 | 1144 | 1232 | 1320 | 1408 | 1496 | 1584 | 1672 | 1760 | **88** |
| **89** | 979 | 1068 | 1157 | 1246 | 1335 | 1424 | 1513 | 1602 | 1691 | 1780 | **89** |
| **90** | 990 | 1080 | 1170 | 1260 | 1350 | 1440 | 1530 | 1620 | 1710 | 1800 | **90** |
| **91** | 1001 | 1092 | 1183 | 1274 | 1365 | 1456 | 1547 | 1638 | 1729 | 1820 | **91** |
| **92** | 1012 | 1104 | 1196 | 1288 | 1380 | 1472 | 1564 | 1656 | 1748 | 1840 | **92** |
| **93** | 1023 | 1116 | 1209 | 1302 | 1395 | 1488 | 1581 | 1674 | 1767 | 1860 | **93** |
| **94** | 1034 | 1128 | 1222 | 1316 | 1410 | 1504 | 1598 | 1692 | 1786 | 1880 | **94** |
| **95** | 1045 | 1140 | 1235 | 1330 | 1425 | 1520 | 1615 | 1710 | 1805 | 1900 | **95** |
| **96** | 1056 | 1152 | 1248 | 1344 | 1440 | 1536 | 1632 | 1728 | 1824 | 1920 | **96** |
| **97** | 1067 | 1164 | 1261 | 1358 | 1455 | 1552 | 1649 | 1746 | 1843 | 1940 | **97** |
| **98** | 1078 | 1176 | 1274 | 1372 | 1470 | 1568 | 1666 | 1764 | 1862 | 1960 | **98** |
| **99** | 1089 | 1188 | 1287 | 1386 | 1485 | 1584 | 1683 | 1782 | 1881 | 1980 | **99** |
| **100** | 1100 | 1200 | 1300 | 1400 | 1500 | 1600 | 1700 | 1800 | 1900 | 2000 | **100** |
| | 11 | 12 | 13 | 14 | 15 | 16 | 17 | 18 | 19 | 20 | |

| | 21 | 22 | 23 | 24 | 25 | 26 | 27 | 28 | 29 | 30 | |
|---|---|---|---|---|---|---|---|---|---|---|---|
| 1 | 21 | 22 | 23 | 24 | 25 | 26 | 27 | 28 | 29 | 30 | 1 |
| 2 | 42 | 44 | 46 | 48 | 50 | 52 | 54 | 56 | 58 | 60 | 2 |
| 3 | 63 | 66 | 69 | 72 | 75 | 78 | 81 | 84 | 87 | 90 | 3 |
| 4 | 84 | 88 | 92 | 96 | 100 | 104 | 108 | 112 | 116 | 120 | 4 |
| 5 | 105 | 110 | 115 | 120 | 125 | 130 | 135 | 140 | 145 | 150 | 5 |
| 6 | 126 | 132 | 138 | 144 | 150 | 156 | 162 | 168 | 174 | 180 | 6 |
| 7 | 147 | 154 | 161 | 168 | 175 | 182 | 189 | 196 | 203 | 210 | 7 |
| 8 | 168 | 176 | 184 | 192 | 200 | 208 | 216 | 224 | 232 | 240 | 8 |
| 9 | 189 | 198 | 207 | 216 | 225 | 234 | 243 | 252 | 261 | 270 | 9 |
| 10 | 210 | 220 | 230 | 240 | 250 | 260 | 270 | 280 | 290 | 300 | 10 |
| 11 | 231 | 242 | 253 | 264 | 275 | 286 | 297 | 308 | 319 | 330 | 11 |
| 12 | 252 | 264 | 276 | 288 | 300 | 312 | 324 | 336 | 348 | 360 | 12 |
| 13 | 273 | 286 | 299 | 312 | 325 | 338 | 351 | 364 | 377 | 390 | 13 |
| 14 | 294 | 308 | 322 | 336 | 350 | 364 | 378 | 392 | 406 | 420 | 14 |
| 15 | 315 | 330 | 345 | 360 | 375 | 390 | 405 | 420 | 435 | 450 | 15 |
| 16 | 336 | 352 | 368 | 384 | 400 | 416 | 432 | 448 | 464 | 480 | 16 |
| 17 | 357 | 374 | 391 | 408 | 425 | 442 | 459 | 476 | 493 | 510 | 17 |
| 18 | 378 | 396 | 414 | 432 | 450 | 468 | 486 | 504 | 522 | 540 | 18 |
| 19 | 399 | 418 | 437 | 456 | 475 | 494 | 513 | 532 | 551 | 570 | 19 |
| 20 | 420 | 440 | 460 | 480 | 500 | 520 | 540 | 560 | 580 | 600 | 20 |
| 21 | 441 | 462 | 483 | 504 | 525 | 546 | 567 | 588 | 609 | 630 | 21 |
| 22 | 462 | 484 | 506 | 528 | 550 | 572 | 594 | 616 | 638 | 660 | 22 |
| 23 | 483 | 506 | 529 | 552 | 575 | 598 | 621 | 644 | 667 | 690 | 23 |
| 24 | 504 | 528 | 552 | 576 | 600 | 624 | 648 | 672 | 696 | 720 | 24 |
| 25 | 525 | 550 | 575 | 600 | 625 | 650 | 675 | 700 | 725 | 750 | 25 |
| 26 | 546 | 572 | 598 | 624 | 650 | 676 | 702 | 728 | 754 | 780 | 26 |
| 27 | 567 | 594 | 621 | 648 | 675 | 702 | 729 | 756 | 783 | 810 | 27 |
| 28 | 588 | 616 | 644 | 672 | 700 | 728 | 756 | 784 | 812 | 840 | 28 |
| 29 | 609 | 638 | 667 | 696 | 725 | 754 | 783 | 812 | 841 | 870 | 29 |
| 30 | 630 | 660 | 690 | 720 | 750 | 780 | 810 | 840 | 870 | 900 | 30 |
| 31 | 651 | 682 | 713 | 744 | 775 | 806 | 837 | 868 | 899 | 930 | 31 |
| 32 | 672 | 704 | 736 | 768 | 800 | 832 | 864 | 896 | 928 | 960 | 32 |
| 33 | 693 | 726 | 759 | 792 | 825 | 858 | 891 | 924 | 957 | 990 | 33 |
| 34 | 714 | 748 | 782 | 816 | 850 | 884 | 918 | 952 | 986 | 1020 | 34 |
| 35 | 735 | 770 | 805 | 840 | 875 | 910 | 945 | 980 | 1015 | 1050 | 35 |
| 36 | 756 | 792 | 828 | 864 | 900 | 936 | 972 | 1008 | 1044 | 1080 | 36 |
| 37 | 777 | 814 | 851 | 888 | 925 | 962 | 999 | 1036 | 1073 | 1110 | 37 |
| 38 | 798 | 836 | 874 | 912 | 950 | 988 | 1026 | 1064 | 1102 | 1140 | 38 |
| 39 | 819 | 858 | 897 | 936 | 975 | 1014 | 1053 | 1092 | 1131 | 1170 | 39 |
| 40 | 840 | 880 | 920 | 960 | 1000 | 1040 | 1080 | 1120 | 1160 | 1200 | 40 |
| 41 | 861 | 902 | 943 | 984 | 1025 | 1066 | 1107 | 1148 | 1189 | 1230 | 42 |
| 42 | 882 | 924 | 966 | 1008 | 1050 | 1092 | 1134 | 1176 | 1218 | 1260 | 41 |
| 43 | 903 | 946 | 989 | 1032 | 1075 | 1118 | 1161 | 1204 | 1247 | 1290 | 43 |
| 44 | 924 | 968 | 1012 | 1056 | 1100 | 1144 | 1188 | 1232 | 1276 | 1320 | 44 |
| 45 | 945 | 990 | 1035 | 1080 | 1125 | 1170 | 1215 | 1260 | 1305 | 1350 | 45 |
| 46 | 966 | 1012 | 1058 | 1104 | 1150 | 1196 | 1242 | 1288 | 1334 | 1380 | 46 |
| 47 | 987 | 1034 | 1081 | 1128 | 1175 | 1222 | 1269 | 1316 | 1363 | 1410 | 47 |
| 48 | 1008 | 1056 | 1104 | 1152 | 1200 | 1248 | 1296 | 1344 | 1392 | 1440 | 48 |
| 49 | 1029 | 1078 | 1127 | 1176 | 1225 | 1274 | 1323 | 1372 | 1421 | 1470 | 49 |
| 50 | 1050 | 1100 | 1150 | 1200 | 1250 | 1300 | 1350 | 1400 | 1450 | 1500 | 50 |
| | 21 | 22 | 23 | 24 | 25 | 26 | 27 | 28 | 29 | 30 | |

| | 21 | 22 | 23 | 24 | 25 | 26 | 27 | 28 | 29 | 30 | |
|---|---|---|---|---|---|---|---|---|---|---|---|
| **51** | 1071 | 1122 | 1173 | 1224 | 1275 | 1326 | 1377 | 1428 | 1479 | 1530 | **51** |
| **52** | 1092 | 1144 | 1196 | 1248 | 1300 | 1352 | 1404 | 1456 | 1508 | 1560 | **52** |
| **53** | 1113 | 1166 | 1219 | 1272 | 1325 | 1378 | 1431 | 1484 | 1537 | 1590 | **53** |
| **54** | 1134 | 1188 | 1242 | 1296 | 1350 | 1404 | 1458 | 1512 | 1566 | 1620 | **54** |
| **55** | 1155 | 1210 | 1265 | 1320 | 1375 | 1430 | 1485 | 1540 | 1595 | 1650 | **55** |
| **56** | 1176 | 1232 | 1288 | 1344 | 1400 | 1456 | 1512 | 1568 | 1624 | 1680 | **56** |
| **57** | 1197 | 1254 | 1311 | 1368 | 1425 | 1482 | 1539 | 1596 | 1653 | 1710 | **57** |
| **58** | 1218 | 1276 | 1334 | 1392 | 1450 | 1508 | 1566 | 1624 | 1682 | 1740 | **58** |
| **59** | 1239 | 1298 | 1357 | 1416 | 1475 | 1534 | 1593 | 1652 | 1711 | 1770 | **59** |
| **60** | 1260 | 1320 | 1380 | 1440 | 1500 | 1560 | 1620 | 1680 | 1740 | 1800 | **60** |
| **61** | 1281 | 1342 | 1403 | 1464 | 1525 | 1586 | 1647 | 1708 | 1769 | 1830 | **61** |
| **62** | 1302 | 1364 | 1426 | 1488 | 1550 | 1612 | 1674 | 1736 | 1798 | 1860 | **62** |
| **63** | 1323 | 1386 | 1449 | 1512 | 1575 | 1638 | 1701 | 1764 | 1827 | 1890 | **63** |
| **64** | 1344 | 1408 | 1472 | 1536 | 1600 | 1664 | 1728 | 1792 | 1856 | 1920 | **64** |
| **65** | 1365 | 1430 | 1495 | 1560 | 1625 | 1690 | 1755 | 1820 | 1885 | 1950 | **65** |
| **66** | 1386 | 1452 | 1518 | 1584 | 1650 | 1716 | 1782 | 1848 | 1914 | 1980 | **66** |
| **67** | 1407 | 1474 | 1541 | 1608 | 1675 | 1742 | 1809 | 1876 | 1943 | 2010 | **67** |
| **68** | 1428 | 1496 | 1564 | 1632 | 1700 | 1768 | 1836 | 1904 | 1972 | 2040 | **68** |
| **69** | 1449 | 1518 | 1587 | 1656 | 1725 | 1794 | 1863 | 1932 | 2001 | 2070 | **69** |
| **70** | 1470 | 1540 | 1610 | 1680 | 1750 | 1820 | 1890 | 1960 | 2030 | 2100 | **70** |
| **71** | 1491 | 1562 | 1633 | 1704 | 1775 | 1846 | 1917 | 1988 | 2059 | 2130 | **71** |
| **72** | 1512 | 1584 | 1656 | 1728 | 1800 | 1872 | 1944 | 2016 | 2088 | 2160 | **72** |
| **73** | 1533 | 1606 | 1679 | 1752 | 1825 | 1898 | 1971 | 2044 | 2117 | 2190 | **73** |
| **74** | 1554 | 1628 | 1702 | 1776 | 1850 | 1924 | 1998 | 2072 | 2146 | 2220 | **74** |
| **75** | 1575 | 1650 | 1725 | 1800 | 1875 | 1950 | 2025 | 2100 | 2175 | 2250 | **75** |
| **76** | 1596 | 1672 | 1748 | 1824 | 1900 | 1976 | 2052 | 2128 | 2204 | 2280 | **76** |
| **77** | 1617 | 1694 | 1771 | 1848 | 1925 | 2002 | 2079 | 2156 | 2233 | 2310 | **77** |
| **78** | 1638 | 1716 | 1794 | 1872 | 1950 | 2028 | 2106 | 2184 | 2262 | 2340 | **78** |
| **79** | 1659 | 1738 | 1817 | 1896 | 1975 | 2054 | 2133 | 2212 | 2291 | 2370 | **79** |
| **80** | 1680 | 1760 | 1840 | 1920 | 2000 | 2080 | 2160 | 2240 | 2320 | 2400 | **80** |
| **81** | 1701 | 1782 | 1863 | 1944 | 2025 | 2106 | 2187 | 2268 | 2349 | 2430 | **81** |
| **82** | 1722 | 1804 | 1886 | 1968 | 2050 | 2132 | 2214 | 2296 | 2378 | 2460 | **82** |
| **83** | 1743 | 1826 | 1909 | 1992 | 2075 | 2158 | 2241 | 2324 | 2407 | 2490 | **83** |
| **84** | 1764 | 1848 | 1932 | 2016 | 2100 | 2184 | 2268 | 2352 | 2436 | 2520 | **84** |
| **85** | 1785 | 1870 | 1955 | 2040 | 2125 | 2210 | 2295 | 2380 | 2465 | 2550 | **85** |
| **86** | 1806 | 1892 | 1978 | 2064 | 2150 | 2236 | 2322 | 2408 | 2494 | 2580 | **86** |
| **87** | 1827 | 1914 | 2001 | 2088 | 2175 | 2262 | 2349 | 2436 | 2523 | 2610 | **87** |
| **88** | 1848 | 1936 | 2024 | 2112 | 2200 | 2288 | 2376 | 2464 | 2552 | 2640 | **88** |
| **89** | 1869 | 1958 | 2047 | 2136 | 2225 | 2314 | 2403 | 2492 | 2581 | 2670 | **89** |
| **90** | 1890 | 1980 | 2070 | 2160 | 2250 | 2340 | 2430 | 2520 | 2610 | 2700 | **90** |
| **91** | 1911 | 2002 | 2093 | 2184 | 2275 | 2366 | 2457 | 2548 | 2639 | 2730 | **91** |
| **92** | 1932 | 2024 | 2116 | 2208 | 2300 | 2392 | 2484 | 2576 | 2668 | 2760 | **92** |
| **93** | 1953 | 2046 | 2139 | 2232 | 2325 | 2418 | 2511 | 2604 | 2697 | 2790 | **93** |
| **94** | 1974 | 2068 | 2162 | 2256 | 2350 | 2444 | 2538 | 2632 | 2726 | 2820 | **94** |
| **95** | 1995 | 2090 | 2185 | 2280 | 2375 | 2470 | 2565 | 2660 | 2755 | 2850 | **95** |
| **96** | 2016 | 2112 | 2208 | 2304 | 2400 | 2496 | 2592 | 2688 | 2784 | 2880 | **96** |
| **97** | 2037 | 2134 | 2231 | 2328 | 2425 | 2522 | 2619 | 2716 | 2813 | 2910 | **97** |
| **98** | 2058 | 2156 | 2254 | 2352 | 2450 | 2548 | 2646 | 2744 | 2842 | 2940 | **98** |
| **99** | 2079 | 2178 | 2277 | 2376 | 2475 | 2574 | 2673 | 2772 | 2871 | 2970 | **99** |
| **100** | 2100 | 2200 | 2300 | 2400 | 2500 | 2600 | 2700 | 2800 | 2900 | 3000 | **100** |
| | **21** | **22** | **23** | **24** | **25** | **26** | **27** | **28** | **29** | **30** | |

| | 31 | 32 | 33 | 34 | 35 | 36 | 37 | 38 | 39 | 40 | |
|---|---|---|---|---|---|---|---|---|---|---|---|
| 1 | 31 | 32 | 33 | 34 | 35 | 36 | 37 | 38 | 39 | 40 | 1 |
| 2 | 62 | 64 | 66 | 68 | 70 | 72 | 74 | 76 | 78 | 80 | 2 |
| 3 | 93 | 96 | 99 | 102 | 105 | 108 | 111 | 114 | 117 | 120 | 3 |
| 4 | 124 | 128 | 132 | 136 | 140 | 144 | 148 | 152 | 156 | 160 | 4 |
| 5 | 155 | 160 | 165 | 170 | 175 | 180 | 185 | 190 | 195 | 200 | 5 |
| 6 | 186 | 192 | 198 | 204 | 210 | 216 | 222 | 228 | 234 | 240 | 6 |
| 7 | 217 | 224 | 231 | 238 | 245 | 252 | 259 | 266 | 273 | 280 | 7 |
| 8 | 248 | 256 | 264 | 272 | 280 | 288 | 296 | 304 | 312 | 320 | 8 |
| 9 | 279 | 288 | 297 | 306 | 315 | 324 | 333 | 342 | 351 | 360 | 9 |
| 10 | 310 | 320 | 330 | 340 | 350 | 360 | 370 | 380 | 390 | 400 | 10 |
| 11 | 341 | 352 | 363 | 374 | 385 | 396 | 407 | 418 | 429 | 440 | 11 |
| 12 | 372 | 384 | 396 | 408 | 420 | 432 | 444 | 456 | 468 | 480 | 12 |
| 13 | 403 | 416 | 429 | 442 | 455 | 468 | 481 | 494 | 507 | 520 | 13 |
| 14 | 434 | 448 | 462 | 476 | 490 | 504 | 518 | 532 | 546 | 560 | 14 |
| 15 | 465 | 480 | 495 | 510 | 525 | 540 | 555 | 570 | 585 | 600 | 15 |
| 16 | 496 | 512 | 528 | 544 | 560 | 576 | 592 | 608 | 624 | 640 | 16 |
| 17 | 527 | 544 | 561 | 578 | 595 | 612 | 629 | 646 | 663 | 680 | 17 |
| 18 | 558 | 576 | 594 | 612 | 630 | 648 | 666 | 684 | 702 | 720 | 18 |
| 19 | 589 | 608 | 627 | 646 | 665 | 684 | 703 | 722 | 741 | 760 | 19 |
| 20 | 620 | 640 | 660 | 680 | 700 | 720 | 740 | 760 | 780 | 800 | 20 |
| 21 | 651 | 672 | 693 | 714 | 735 | 756 | 777 | 798 | 819 | 840 | 21 |
| 22 | 682 | 704 | 726 | 748 | 770 | 792 | 814 | 836 | 858 | 880 | 22 |
| 23 | 713 | 736 | 759 | 782 | 805 | 828 | 851 | 874 | 897 | 920 | 23 |
| 24 | 744 | 768 | 792 | 816 | 840 | 864 | 888 | 912 | 936 | 960 | 24 |
| 25 | 775 | 800 | 825 | 850 | 875 | 900 | 925 | 950 | 975 | 1000 | 25 |
| 26 | 806 | 832 | 858 | 884 | 910 | 936 | 962 | 988 | 1014 | 1040 | 26 |
| 27 | 837 | 864 | 891 | 918 | 945 | 972 | 999 | 1026 | 1053 | 1080 | 27 |
| 28 | 868 | 896 | 924 | 952 | 980 | 1008 | 1036 | 1064 | 1092 | 1120 | 28 |
| 29 | 899 | 928 | 957 | 986 | 1015 | 1044 | 1073 | 1102 | 1131 | 1160 | 29 |
| 30 | 930 | 960 | 990 | 1020 | 1050 | 1080 | 1110 | 1140 | 1170 | 1200 | 30 |
| 31 | 961 | 992 | 1023 | 1054 | 1085 | 1116 | 1147 | 1178 | 1209 | 1240 | 31 |
| 32 | 992 | 1024 | 1056 | 1088 | 1120 | 1152 | 1184 | 1216 | 1248 | 1280 | 32 |
| 33 | 1023 | 1056 | 1089 | 1122 | 1155 | 1188 | 1221 | 1254 | 1287 | 1320 | 33 |
| 34 | 1054 | 1088 | 1122 | 1156 | 1190 | 1224 | 1258 | 1292 | 1326 | 1360 | 34 |
| 35 | 1085 | 1120 | 1155 | 1190 | 1225 | 1260 | 1295 | 1330 | 1365 | 1400 | 35 |
| 36 | 1116 | 1152 | 1188 | 1224 | 1260 | 1296 | 1332 | 1368 | 1404 | 1440 | 36 |
| 37 | 1147 | 1184 | 1221 | 1258 | 1295 | 1332 | 1369 | 1406 | 1443 | 1480 | 37 |
| 38 | 1178 | 1216 | 1254 | 1292 | 1330 | 1368 | 1406 | 1444 | 1482 | 1520 | 38 |
| 39 | 1209 | 1248 | 1287 | 1326 | 1365 | 1404 | 1443 | 1482 | 1521 | 1560 | 39 |
| 40 | 1240 | 1280 | 1320 | 1360 | 1400 | 1440 | 1480 | 1520 | 1560 | 1600 | 40 |
| 41 | 1271 | 1312 | 1353 | 1394 | 1435 | 1476 | 1517 | 1558 | 1599 | 1640 | 41 |
| 42 | 1302 | 1344 | 1386 | 1428 | 1470 | 1512 | 1554 | 1596 | 1638 | 1680 | 42 |
| 43 | 1333 | 1376 | 1419 | 1462 | 1505 | 1548 | 1591 | 1634 | 1677 | 1720 | 43 |
| 44 | 1364 | 1408 | 1452 | 1496 | 1540 | 1584 | 1628 | 1672 | 1716 | 1760 | 44 |
| 45 | 1395 | 1440 | 1485 | 1530 | 1575 | 1620 | 1665 | 1710 | 1755 | 1800 | 45 |
| 46 | 1426 | 1472 | 1518 | 1564 | 1610 | 1656 | 1702 | 1748 | 1794 | 1840 | 46 |
| 47 | 1457 | 1504 | 1551 | 1598 | 1645 | 1692 | 1739 | 1786 | 1833 | 1880 | 47 |
| 48 | 1488 | 1536 | 1584 | 1632 | 1680 | 1728 | 1776 | 1824 | 1872 | 1920 | 48 |
| 49 | 1519 | 1568 | 1617 | 1666 | 1715 | 1764 | 1813 | 1862 | 1911 | 1960 | 49 |
| 50 | 1550 | 1600 | 1650 | 1700 | 1750 | 1800 | 1850 | 1900 | 1950 | 2000 | 50 |
| | 31 | 32 | 33 | 34 | 35 | 36 | 37 | 38 | 39 | 40 | |

| | 31 | 32 | 33 | 34 | 35 | 36 | 37 | 38 | 39 | 40 | |
|---|---|---|---|---|---|---|---|---|---|---|---|
| **51** | 1581 | 1632 | 1683 | 1734 | 1785 | 1836 | 1887 | 1938 | 1989 | 2040 | **51** |
| **52** | 1612 | 1664 | 1716 | 1768 | 1820 | 1872 | 1924 | 1976 | 2028 | 2080 | **52** |
| **53** | 1643 | 1696 | 1749 | 1802 | 1855 | 1908 | 1961 | 2014 | 2067 | 2120 | **53** |
| **54** | 1674 | 1728 | 1782 | 1836 | 1890 | 1944 | 1998 | 2052 | 2106 | 2160 | **54** |
| **55** | 1705 | 1760 | 1815 | 1870 | 1925 | 1980 | 2035 | 2090 | 2145 | 2200 | **55** |
| **56** | 1736 | 1792 | 1848 | 1904 | 1960 | 2016 | 2072 | 2128 | 2184 | 2240 | **56** |
| **57** | 1767 | 1824 | 1881 | 1938 | 1995 | 2052 | 2109 | 2166 | 2223 | 2280 | **57** |
| **58** | 1798 | 1856 | 1914 | 1972 | 2030 | 2088 | 2146 | 2204 | 2262 | 2320 | **58** |
| **59** | 1829 | 1888 | 1947 | 2006 | 2065 | 2124 | 2183 | 2242 | 2301 | 2360 | **59** |
| **60** | 1860 | 1920 | 1980 | 2040 | 2100 | 2160 | 2220 | 2280 | 2340 | 2400 | **60** |
| **61** | 1891 | 1952 | 2013 | 2074 | 2135 | 2196 | 2257 | 2318 | 2379 | 2440 | **61** |
| **62** | 1922 | 1984 | 2046 | 2108 | 2170 | 2232 | 2294 | 2356 | 2418 | 2480 | **62** |
| **63** | 1953 | 2016 | 2079 | 2142 | 2205 | 2268 | 2331 | 2394 | 2457 | 2520 | **63** |
| **64** | 1984 | 2048 | 2112 | 2176 | 2240 | 2304 | 2368 | 2432 | 2496 | 2560 | **64** |
| **65** | 2015 | 2080 | 2145 | 2210 | 2275 | 2340 | 2405 | 2470 | 2535 | 2600 | **65** |
| **66** | 2046 | 2112 | 2178 | 2244 | 2310 | 2376 | 2442 | 2508 | 2574 | 2640 | **66** |
| **67** | 2077 | 2144 | 2211 | 2278 | 2345 | 2412 | 2479 | 2546 | 2613 | 2680 | **67** |
| **68** | 2108 | 2176 | 2244 | 2312 | 2380 | 2448 | 2516 | 2584 | 2652 | 2720 | **68** |
| **69** | 2139 | 2208 | 2277 | 2346 | 2415 | 2484 | 2553 | 2622 | 2691 | 2760 | **69** |
| **70** | 2170 | 2240 | 2310 | 2380 | 2450 | 2520 | 2590 | 2660 | 2730 | 2800 | **70** |
| **71** | 2201 | 2272 | 2343 | 2414 | 2485 | 2556 | 2627 | 2698 | 2769 | 2840 | **71** |
| **72** | 2232 | 2304 | 2376 | 2448 | 2520 | 2592 | 2664 | 2736 | 2808 | 2880 | **72** |
| **73** | 2263 | 2336 | 2409 | 2482 | 2555 | 2628 | 2701 | 2774 | 2847 | 2920 | **73** |
| **74** | 2294 | 2368 | 2442 | 2516 | 2590 | 2664 | 2738 | 2812 | 2886 | 2960 | **74** |
| **75** | 2325 | 2400 | 2475 | 2550 | 2625 | 2700 | 2775 | 2850 | 2925 | 3000 | **75** |
| **76** | 2356 | 2432 | 2508 | 2584 | 2660 | 2736 | 2812 | 2888 | 2964 | 3040 | **76** |
| **77** | 2387 | 2464 | 2541 | 2618 | 2695 | 2772 | 2849 | 2926 | 3003 | 3080 | **77** |
| **78** | 2418 | 2496 | 2574 | 2652 | 2730 | 2808 | 2886 | 2964 | 3042 | 3120 | **78** |
| **79** | 2449 | 2528 | 2607 | 2686 | 2765 | 2844 | 2923 | 3002 | 3081 | 3160 | **79** |
| **80** | 2480 | 2560 | 2640 | 2720 | 2800 | 2880 | 2960 | 3040 | 3120 | 3200 | **80** |
| **81** | 2511 | 2592 | 2673 | 2754 | 2835 | 2916 | 2997 | 3078 | 3159 | 3240 | **81** |
| **82** | 2542 | 2624 | 2706 | 2788 | 2870 | 2952 | 3034 | 3116 | 3198 | 3280 | **82** |
| **83** | 2573 | 2656 | 2739 | 2822 | 2905 | 2988 | 3071 | 3154 | 3237 | 3320 | **83** |
| **84** | 2604 | 2688 | 2772 | 2856 | 2940 | 3024 | 3108 | 3192 | 3276 | 3360 | **84** |
| **85** | 2635 | 2720 | 2805 | 2890 | 2975 | 3060 | 3145 | 3230 | 3315 | 3400 | **85** |
| **86** | 2666 | 2752 | 2838 | 2924 | 3010 | 3096 | 3182 | 3268 | 3354 | 3440 | **86** |
| **87** | 2697 | 2784 | 2871 | 2958 | 3045 | 3132 | 3219 | 3306 | 3393 | 3480 | **87** |
| **88** | 2728 | 2816 | 2904 | 2992 | 3080 | 3168 | 3256 | 3344 | 3432 | 3520 | **88** |
| **89** | 2759 | 2848 | 2937 | 3026 | 3115 | 3204 | 3293 | 3382 | 3471 | 3560 | **89** |
| **90** | 2790 | 2880 | 2970 | 3060 | 3150 | 3240 | 3330 | 3420 | 3510 | 3600 | **90** |
| **91** | 2821 | 2912 | 3003 | 3094 | 3185 | 3276 | 3367 | 3458 | 3549 | 3640 | **91** |
| **92** | 2852 | 2944 | 3036 | 3128 | 3220 | 3312 | 3404 | 3496 | 3588 | 3680 | **92** |
| **93** | 2883 | 2976 | 3069 | 3162 | 3255 | 3348 | 3441 | 3534 | 3627 | 3720 | **93** |
| **94** | 2914 | 3008 | 3102 | 3196 | 3290 | 3384 | 3478 | 3572 | 3666 | 3760 | **94** |
| **95** | 2945 | 3040 | 3135 | 3230 | 3325 | 3420 | 3515 | 3610 | 3705 | 3800 | **95** |
| **96** | 2976 | 3072 | 3168 | 3264 | 3360 | 3456 | 3552 | 3648 | 3744 | 3840 | **96** |
| **97** | 3007 | 3104 | 3201 | 3298 | 3395 | 3492 | 3589 | 3686 | 3783 | 3880 | **97** |
| **98** | 3038 | 3136 | 3234 | 3332 | 3430 | 3528 | 3626 | 3724 | 3822 | 3920 | **98** |
| **99** | 3069 | 3168 | 3267 | 3366 | 3465 | 3564 | 3663 | 3762 | 3861 | 3960 | **99** |
| **100** | 3100 | 3200 | 3300 | 3400 | 3500 | 3600 | 3700 | 3800 | 3900 | 4000 | **100** |
| | **31** | **32** | **33** | **34** | **35** | **36** | **37** | **38** | **39** | **40** | |

| | 41 | 42 | 43 | 44 | 45 | 46 | 47 | 48 | 49 | 50 | |
|---|---|---|---|---|---|---|---|---|---|---|---|
| 1 | 41 | 42 | 43 | 44 | 45 | 46 | 47 | 48 | 49 | 50 | 1 |
| 2 | 82 | 84 | 86 | 88 | 90 | 92 | 94 | 96 | 98 | 100 | 2 |
| 3 | 123 | 126 | 129 | 132 | 135 | 138 | 141 | 144 | 147 | 150 | 3 |
| 4 | 164 | 168 | 172 | 176 | 180 | 184 | 188 | 192 | 196 | 200 | 4 |
| 5 | 205 | 210 | 215 | 220 | 225 | 230 | 235 | 240 | 245 | 250 | 5 |
| 6 | 246 | 252 | 258 | 264 | 270 | 276 | 282 | 288 | 294 | 300 | 6 |
| 7 | 287 | 294 | 301 | 308 | 315 | 322 | 329 | 336 | 343 | 350 | 7 |
| 8 | 328 | 336 | 344 | 352 | 360 | 368 | 376 | 384 | 392 | 400 | 8 |
| 9 | 369 | 378 | 387 | 396 | 405 | 414 | 423 | 432 | 441 | 450 | 9 |
| 10 | 410 | 420 | 430 | 440 | 450 | 460 | 470 | 480 | 490 | 500 | 10 |
| 11 | 451 | 462 | 473 | 484 | 495 | 506 | 517 | 528 | 539 | 550 | 11 |
| 12 | 492 | 504 | 516 | 528 | 540 | 552 | 564 | 576 | 588 | 600 | 12 |
| 13 | 533 | 546 | 559 | 572 | 585 | 598 | 611 | 624 | 637 | 650 | 13 |
| 14 | 574 | 588 | 602 | 616 | 630 | 644 | 658 | 672 | 686 | 700 | 14 |
| 15 | 615 | 630 | 645 | 660 | 675 | 690 | 705 | 720 | 735 | 750 | 15 |
| 16 | 656 | 672 | 688 | 704 | 720 | 736 | 752 | 768 | 784 | 800 | 16 |
| 17 | 697 | 714 | 731 | 748 | 765 | 782 | 799 | 816 | 833 | 850 | 17 |
| 18 | 738 | 756 | 774 | 792 | 810 | 828 | 846 | 864 | 882 | 900 | 18 |
| 19 | 779 | 798 | 817 | 836 | 855 | 874 | 893 | 912 | 931 | 950 | 19 |
| 20 | 820 | 840 | 860 | 880 | 900 | 920 | 940 | 960 | 980 | 1000 | 20 |
| 21 | 861 | 882 | 903 | 924 | 945 | 966 | 987 | 1008 | 1029 | 1050 | 21 |
| 22 | 902 | 924 | 946 | 968 | 990 | 1012 | 1034 | 1056 | 1078 | 1100 | 22 |
| 23 | 943 | 966 | 989 | 1012 | 1035 | 1058 | 1081 | 1104 | 1127 | 1150 | 23 |
| 24 | 984 | 1008 | 1032 | 1056 | 1080 | 1104 | 1128 | 1152 | 1176 | 1200 | 24 |
| 25 | 1025 | 1050 | 1075 | 1100 | 1125 | 1150 | 1175 | 1200 | 1225 | 1250 | 25 |
| 26 | 1066 | 1092 | 1118 | 1144 | 1170 | 1196 | 1222 | 1248 | 1274 | 1300 | 26 |
| 27 | 1107 | 1134 | 1161 | 1188 | 1215 | 1242 | 1269 | 1296 | 1323 | 1350 | 27 |
| 28 | 1148 | 1176 | 1204 | 1232 | 1260 | 1288 | 1316 | 1344 | 1372 | 1400 | 28 |
| 29 | 1189 | 1218 | 1247 | 1276 | 1305 | 1334 | 1363 | 1392 | 1421 | 1450 | 29 |
| 30 | 1230 | 1260 | 1290 | 1320 | 1350 | 1380 | 1410 | 1440 | 1470 | 1500 | 30 |
| 31 | 1271 | 1302 | 1333 | 1364 | 1395 | 1426 | 1457 | 1488 | 1519 | 1550 | 31 |
| 32 | 1312 | 1344 | 1376 | 1408 | 1440 | 1472 | 1504 | 1536 | 1568 | 1600 | 32 |
| 33 | 1353 | 1386 | 1419 | 1452 | 1485 | 1518 | 1551 | 1584 | 1617 | 1650 | 33 |
| 34 | 1394 | 1428 | 1462 | 1496 | 1530 | 1564 | 1598 | 1632 | 1666 | 1700 | 34 |
| 35 | 1435 | 1470 | 1505 | 1540 | 1575 | 1610 | 1645 | 1680 | 1715 | 1750 | 35 |
| 36 | 1476 | 1512 | 1548 | 1584 | 1620 | 1656 | 1692 | 1728 | 1764 | 1800 | 36 |
| 37 | 1517 | 1554 | 1591 | 1628 | 1665 | 1702 | 1739 | 1776 | 1813 | 1850 | 37 |
| 38 | 1558 | 1596 | 1634 | 1672 | 1710 | 1748 | 1786 | 1824 | 1862 | 1900 | 38 |
| 39 | 1599 | 1638 | 1677 | 1716 | 1755 | 1794 | 1833 | 1872 | 1911 | 1950 | 39 |
| 40 | 1640 | 1680 | 1720 | 1760 | 1800 | 1840 | 1880 | 1920 | 1960 | 2000 | 40 |
| 41 | 1681 | 1722 | 1763 | 1804 | 1845 | 1886 | 1927 | 1968 | 2009 | 2050 | 41 |
| 42 | 1722 | 1764 | 1806 | 1848 | 1890 | 1932 | 1974 | 2016 | 2058 | 2100 | 42 |
| 43 | 1763 | 1806 | 1849 | 1892 | 1935 | 1978 | 2021 | 2064 | 2107 | 2150 | 43 |
| 44 | 1804 | 1848 | 1892 | 1936 | 1980 | 2024 | 2068 | 2112 | 2156 | 2200 | 44 |
| 45 | 1845 | 1890 | 1935 | 1980 | 2025 | 2070 | 2115 | 2160 | 2205 | 2250 | 45 |
| 46 | 1886 | 1932 | 1978 | 2024 | 2070 | 2116 | 2162 | 2208 | 2254 | 2300 | 46 |
| 47 | 1927 | 1974 | 2021 | 2068 | 2115 | 2162 | 2209 | 2256 | 2303 | 2350 | 47 |
| 48 | 1968 | 2016 | 2064 | 2112 | 2160 | 2208 | 2256 | 2304 | 2352 | 2400 | 48 |
| 49 | 2009 | 2058 | 2107 | 2156 | 2205 | 2254 | 2303 | 2352 | 2401 | 2450 | 49 |
| 50 | 2050 | 2100 | 2150 | 2200 | 2250 | 2300 | 2350 | 2400 | 2450 | 2500 | 50 |
| | 41 | 42 | 43 | 44 | 45 | 46 | 47 | 48 | 49 | 50 | |

| | 41 | 42 | 43 | 44 | 45 | 46 | 47 | 48 | 49 | 50 | |
|---|---|---|---|---|---|---|---|---|---|---|---|
| **51** | 2091 | 2142 | 2193 | 2244 | 2295 | 2346 | 2397 | 2448 | 2499 | 2550 | **51** |
| **52** | 2132 | 2184 | 2236 | 2288 | 2340 | 2392 | 2444 | 2496 | 2548 | 2600 | **52** |
| **53** | 2173 | 2226 | 2279 | 2332 | 2385 | 2438 | 2491 | 2544 | 2597 | 2650 | **53** |
| **54** | 2214 | 2268 | 2322 | 2376 | 2430 | 2484 | 2538 | 2592 | 2646 | 2700 | **54** |
| **55** | 2255 | 2310 | 2365 | 2420 | 2475 | 2530 | 2585 | 2640 | 2695 | 2750 | **55** |
| **56** | 2296 | 2352 | 2408 | 2464 | 2520 | 2576 | 2632 | 2688 | 2744 | 2800 | **56** |
| **57** | 2337 | 2394 | 2451 | 2508 | 2565 | 2622 | 2679 | 2736 | 2793 | 2850 | **57** |
| **58** | 2378 | 2436 | 2494 | 2552 | 2610 | 2668 | 2726 | 2784 | 2842 | 2900 | **58** |
| **59** | 2419 | 2478 | 2537 | 2596 | 2655 | 2714 | 2773 | 2832 | 2891 | 2950 | **59** |
| **60** | 2460 | 2520 | 2580 | 2640 | 2700 | 2760 | 2820 | 2880 | 2940 | 3000 | **60** |
| **61** | 2501 | 2562 | 2623 | 2684 | 2745 | 2806 | 2867 | 2928 | 2989 | 3050 | **61** |
| **62** | 2542 | 2604 | 2666 | 2728 | 2790 | 2852 | 2914 | 2976 | 3038 | 3100 | **62** |
| **63** | 2583 | 2646 | 2709 | 2772 | 2835 | 2898 | 2961 | 3024 | 3087 | 3150 | **63** |
| **64** | 2624 | 2688 | 2752 | 2816 | 2880 | 2944 | 3008 | 3072 | 3136 | 3200 | **64** |
| **65** | 2665 | 2730 | 2795 | 2860 | 2925 | 2990 | 3055 | 3120 | 3185 | 3250 | **65** |
| **66** | 2706 | 2772 | 2838 | 2904 | 2970 | 3036 | 3102 | 3168 | 3234 | 3300 | **66** |
| **67** | 2747 | 2814 | 2881 | 2948 | 3015 | 3082 | 3149 | 3216 | 3283 | 3350 | **67** |
| **68** | 2788 | 2856 | 2924 | 2992 | 3060 | 3128 | 3196 | 3264 | 3332 | 3400 | **68** |
| **69** | 2829 | 2898 | 2967 | 3036 | 3105 | 3174 | 3243 | 3312 | 3381 | 3450 | **69** |
| **70** | 2870 | 2940 | 3010 | 3080 | 3150 | 3220 | 3290 | 3360 | 3430 | 3500 | **07** |
| **71** | 2911 | 2982 | 3053 | 3124 | 3195 | 3266 | 3337 | 3408 | 3479 | 3550 | **71** |
| **72** | 2952 | 3024 | 3096 | 3168 | 3240 | 3312 | 3384 | 3456 | 3528 | 3600 | **72** |
| **73** | 2993 | 3066 | 3139 | 3212 | 3285 | 3358 | 3431 | 3504 | 3577 | 3650 | **73** |
| **74** | 3034 | 3108 | 3182 | 3256 | 3330 | 3404 | 3478 | 3552 | 3626 | 3700 | **74** |
| **75** | 3075 | 3150 | 3225 | 3300 | 3375 | 3450 | 3525 | 3600 | 3675 | 3750 | **75** |
| **76** | 3116 | 3192 | 3268 | 3344 | 3420 | 3496 | 3572 | 3648 | 3724 | 3800 | **76** |
| **77** | 3157 | 3234 | 3311 | 3388 | 3465 | 3542 | 3619 | 3696 | 3773 | 3850 | **77** |
| **78** | 3198 | 3276 | 3354 | 3432 | 3510 | 3588 | 3666 | 3744 | 3822 | 3900 | **78** |
| **79** | 3239 | 3318 | 3397 | 3476 | 3555 | 3634 | 3713 | 3792 | 3871 | 3950 | **79** |
| **80** | 3280 | 3360 | 3440 | 3520 | 3600 | 3680 | 3760 | 3840 | 3920 | 4000 | **80** |
| **81** | 3321 | 3402 | 3483 | 3564 | 3645 | 3726 | 3807 | 3888 | 3969 | 4050 | **81** |
| **82** | 3362 | 3444 | 3526 | 3608 | 3690 | 3772 | 3854 | 3936 | 4018 | 4100 | **82** |
| **83** | 3403 | 3486 | 3569 | 3652 | 3735 | 3818 | 3901 | 3984 | 4067 | 4150 | **83** |
| **84** | 3444 | 3528 | 3612 | 3696 | 3780 | 3864 | 3948 | 4032 | 4116 | 4200 | **84** |
| **85** | 3485 | 3570 | 3655 | 3740 | 3825 | 3910 | 3995 | 4080 | 4165 | 4250 | **85** |
| **86** | 3526 | 3612 | 3698 | 3784 | 3870 | 3956 | 4042 | 4128 | 4214 | 4300 | **86** |
| **87** | 3567 | 3654 | 3741 | 3828 | 3915 | 4002 | 4089 | 4176 | 4263 | 4350 | **87** |
| **88** | 3608 | 3696 | 3784 | 3872 | 3960 | 4048 | 4136 | 4224 | 4312 | 4400 | **88** |
| **89** | 3649 | 3738 | 3827 | 3916 | 4005 | 4094 | 4183 | 4272 | 4361 | 4450 | **89** |
| **90** | 3690 | 3780 | 3870 | 3960 | 4050 | 4140 | 4230 | 4320 | 4410 | 4500 | **90** |
| **91** | 3731 | 3822 | 3913 | 4004 | 4095 | 4186 | 4277 | 4368 | 4459 | 4550 | **91** |
| **92** | 3772 | 3864 | 3956 | 4048 | 4140 | 4232 | 4324 | 4416 | 4508 | 4600 | **92** |
| **93** | 3813 | 3906 | 3999 | 4092 | 4185 | 4278 | 4371 | 4464 | 4557 | 4650 | **93** |
| **94** | 3854 | 3948 | 4042 | 4136 | 4230 | 4324 | 4418 | 4512 | 4606 | 4700 | **94** |
| **95** | 3895 | 3990 | 4085 | 4180 | 4275 | 4370 | 4465 | 4560 | 4655 | 4750 | **95** |
| **96** | 3936 | 4032 | 4128 | 4224 | 4320 | 4416 | 4512 | 4608 | 4704 | 4800 | **96** |
| **97** | 3977 | 4074 | 4171 | 4268 | 4365 | 4462 | 4559 | 4656 | 4753 | 4850 | **97** |
| **98** | 4018 | 4116 | 4214 | 4312 | 4410 | 4508 | 4606 | 4704 | 4802 | 4900 | **98** |
| **99** | 4059 | 4158 | 4257 | 4356 | 4455 | 4554 | 4653 | 4752 | 4851 | 4950 | **99** |
| **100** | 4100 | 4200 | 4300 | 4400 | 4500 | 4600 | 4700 | 4800 | 4900 | 5000 | **100** |
| | 41 | 42 | 43 | 44 | 45 | 46 | 47 | 48 | 49 | 50 | |

| | 51 | 52 | 53 | 54 | 55 | 56 | 57 | 58 | 59 | 60 | |
|---|---|---|---|---|---|---|---|---|---|---|---|
| **1** | 51 | 52 | 53 | 54 | 55 | 56 | 57 | 58 | 59 | 60 | **1** |
| **2** | 102 | 104 | 106 | 108 | 110 | 112 | 114 | 116 | 118 | 120 | **2** |
| **3** | 153 | 156 | 159 | 162 | 165 | 168 | 171 | 174 | 177 | 180 | **3** |
| **4** | 204 | 208 | 212 | 216 | 220 | 224 | 228 | 232 | 236 | 240 | **4** |
| **5** | 255 | 260 | 265 | 270 | 275 | 280 | 285 | 290 | 295 | 300 | **5** |
| **6** | 306 | 312 | 318 | 324 | 330 | 336 | 342 | 348 | 354 | 360 | **6** |
| **7** | 357 | 364 | 371 | 378 | 385 | 392 | 399 | 406 | 413 | 420 | **7** |
| **8** | 408 | 416 | 424 | 432 | 440 | 448 | 456 | 464 | 472 | 480 | **8** |
| **9** | 459 | 468 | 477 | 486 | 495 | 504 | 513 | 522 | 531 | 540 | **9** |
| **10** | 510 | 520 | 530 | 540 | 550 | 560 | 570 | 580 | 590 | 600 | **10** |
| **11** | 561 | 572 | 583 | 594 | 605 | 616 | 627 | 638 | 649 | 660 | **11** |
| **12** | 612 | 624 | 636 | 648 | 660 | 672 | 684 | 696 | 708 | 720 | **12** |
| **13** | 663 | 676 | 689 | 702 | 715 | 728 | 741 | 754 | 767 | 780 | **13** |
| **14** | 714 | 728 | 742 | 756 | 770 | 784 | 798 | 812 | 826 | 840 | **14** |
| **15** | 765 | 780 | 795 | 810 | 825 | 840 | 855 | 870 | 885 | 900 | **15** |
| **16** | 816 | 832 | 848 | 864 | 880 | 896 | 912 | 928 | 944 | 960 | **16** |
| **17** | 867 | 884 | 901 | 918 | 935 | 952 | 969 | 986 | 1003 | 1020 | **17** |
| **18** | 918 | 936 | 954 | 972 | 990 | 1008 | 1026 | 1044 | 1062 | 1080 | **18** |
| **19** | 969 | 988 | 1007 | 1026 | 1045 | 1064 | 1083 | 1102 | 1121 | 1140 | **19** |
| **20** | 1020 | 1040 | 1060 | 1080 | 1100 | 1120 | 1140 | 1160 | 1180 | 1200 | **20** |
| **21** | 1071 | 1092 | 1113 | 1134 | 1155 | 1176 | 1197 | 1218 | 1239 | 1260 | **21** |
| **22** | 1122 | 1144 | 1166 | 1188 | 1210 | 1232 | 1254 | 1276 | 1298 | 1320 | **22** |
| **23** | 1173 | 1196 | 1219 | 1242 | 1265 | 1288 | 1311 | 1334 | 1357 | 1380 | **23** |
| **24** | 1224 | 1248 | 1272 | 1296 | 1320 | 1344 | 1368 | 1392 | 1416 | 1440 | **24** |
| **25** | 1275 | 1300 | 1325 | 1350 | 1375 | 1400 | 1425 | 1450 | 1475 | 1500 | **25** |
| **26** | 1326 | 1352 | 1378 | 1404 | 1430 | 1456 | 1482 | 1508 | 1534 | 1560 | **26** |
| **27** | 1377 | 1404 | 1431 | 1458 | 1485 | 1512 | 1539 | 1566 | 1593 | 1620 | **27** |
| **28** | 1428 | 1456 | 1484 | 1512 | 1540 | 1568 | 1596 | 1624 | 1652 | 1680 | **28** |
| **29** | 1479 | 1508 | 1537 | 1566 | 1595 | 1624 | 1653 | 1682 | 1711 | 1740 | **29** |
| **30** | 1530 | 1560 | 1590 | 1620 | 1650 | 1680 | 1710 | 1740 | 1770 | 1800 | **30** |
| **31** | 1581 | 1612 | 1643 | 1674 | 1705 | 1736 | 1767 | 1798 | 1829 | 1860 | **31** |
| **32** | 1632 | 1664 | 1696 | 1728 | 1760 | 1792 | 1824 | 1856 | 1888 | 1920 | **32** |
| **33** | 1683 | 1716 | 1749 | 1782 | 1815 | 1848 | 1881 | 1914 | 1947 | 1980 | **33** |
| **34** | 1734 | 1768 | 1802 | 1836 | 1870 | 1904 | 1938 | 1972 | 2006 | 2040 | **34** |
| **35** | 1785 | 1820 | 1855 | 1890 | 1925 | 1960 | 1995 | 2030 | 2065 | 2100 | **35** |
| **36** | 1836 | 1872 | 1908 | 1944 | 1980 | 2016 | 2052 | 2088 | 2124 | 2160 | **36** |
| **37** | 1887 | 1924 | 1961 | 1998 | 2035 | 2072 | 2109 | 2146 | 2183 | 2220 | **37** |
| **38** | 1938 | 1976 | 2014 | 2052 | 2090 | 2128 | 2166 | 2204 | 2242 | 2280 | **38** |
| **39** | 1989 | 2028 | 2067 | 2106 | 2145 | 2184 | 2223 | 2262 | 2301 | 2340 | **39** |
| **40** | 2040 | 2080 | 2120 | 2160 | 2200 | 2240 | 2280 | 2320 | 2360 | 2400 | **40** |
| **41** | 2091 | 2132 | 2173 | 2214 | 2255 | 2296 | 2337 | 2378 | 2419 | 2460 | **41** |
| **42** | 2142 | 2184 | 2226 | 2268 | 2310 | 2352 | 2394 | 2436 | 2478 | 2520 | **42** |
| **43** | 2193 | 2236 | 2279 | 2322 | 2365 | 2408 | 2451 | 2494 | 2537 | 2580 | **43** |
| **44** | 2244 | 2288 | 2332 | 2376 | 2420 | 2464 | 2508 | 2552 | 2596 | 2640 | **44** |
| **45** | 2295 | 2340 | 2385 | 2430 | 2475 | 2520 | 2565 | 2610 | 2655 | 2700 | **45** |
| **46** | 2346 | 2392 | 2438 | 2484 | 2530 | 2576 | 2622 | 2668 | 2714 | 2760 | **46** |
| **47** | 2397 | 2444 | 2491 | 2538 | 2585 | 2632 | 2679 | 2726 | 2773 | 2820 | **47** |
| **48** | 2448 | 2496 | 2544 | 2592 | 2640 | 2688 | 2736 | 2784 | 2832 | 2880 | **48** |
| **49** | 2499 | 2548 | 2597 | 2646 | 2695 | 2744 | 2793 | 2842 | 2891 | 2940 | **49** |
| **50** | 2550 | 2600 | 2650 | 2700 | 2750 | 2800 | 2850 | 2900 | 2950 | 3000 | **50** |
| | **51** | **52** | **53** | **54** | **55** | **56** | **57** | **58** | **59** | **60** | |

| | 51 | 52 | 53 | 54 | 55 | 56 | 57 | 58 | 59 | 60 | |
|---|---|---|---|---|---|---|---|---|---|---|---|
| **51** | 2601 | 2652 | 2703 | 2754 | 2805 | 2856 | 2907 | 2958 | 3009 | 3060 | **51** |
| **52** | 2652 | 2704 | 2756 | 2808 | 2860 | 2912 | 2964 | 3016 | 3068 | 3120 | **52** |
| **53** | 2703 | 2756 | 2809 | 2862 | 2915 | 2968 | 3021 | 3074 | 3127 | 3180 | **53** |
| **54** | 2754 | 2808 | 2862 | 2916 | 2970 | 3024 | 3078 | 3132 | 3186 | 3240 | **54** |
| **55** | 2805 | 2860 | 2915 | 2970 | 3025 | 3080 | 3135 | 3190 | 3245 | 3300 | **55** |
| **56** | 2856 | 2912 | 2968 | 3024 | 3080 | 3136 | 3192 | 3248 | 3304 | 3360 | **56** |
| **57** | 2907 | 2964 | 3021 | 3078 | 3135 | 3192 | 3249 | 3306 | 3363 | 3420 | **57** |
| **58** | 2958 | 3016 | 3074 | 3132 | 3190 | 3248 | 3306 | 3364 | 3422 | 3480 | **58** |
| **59** | 3009 | 3068 | 3127 | 3186 | 3245 | 3304 | 3363 | 3422 | 3481 | 3540 | **59** |
| **60** | 3060 | 3120 | 3180 | 3240 | 3300 | 3360 | 3420 | 3480 | 3540 | 3600 | **60** |
| **61** | 3111 | 3172 | 3233 | 3294 | 3355 | 3416 | 3477 | 3538 | 3599 | 3660 | **61** |
| **62** | 3162 | 3224 | 3286 | 3348 | 3410 | 3472 | 3534 | 3596 | 3658 | 3720 | **62** |
| **63** | 3213 | 3276 | 3339 | 3402 | 3465 | 3528 | 3591 | 3654 | 3717 | 3780 | **63** |
| **64** | 3264 | 3328 | 3392 | 3456 | 3520 | 3584 | 3648 | 3712 | 3776 | 3840 | **64** |
| **65** | 3315 | 3380 | 3445 | 3510 | 3575 | 3640 | 3705 | 3770 | 3835 | 3900 | **65** |
| **66** | 3366 | 3432 | 3498 | 3564 | 3630 | 3696 | 3762 | 3828 | 3894 | 3960 | **66** |
| **67** | 3417 | 3484 | 3551 | 3618 | 3685 | 3752 | 3819 | 3886 | 3953 | 4020 | **67** |
| **68** | 3468 | 3536 | 3604 | 3672 | 3740 | 3808 | 3876 | 3944 | 4012 | 4080 | **68** |
| **69** | 3519 | 3588 | 3657 | 3726 | 3795 | 3864 | 3933 | 4002 | 4071 | 4140 | **69** |
| **70** | 3570 | 3640 | 3710 | 3780 | 3850 | 3920 | 3990 | 4060 | 4130 | 4200 | **70** |
| **71** | 3621 | 3692 | 3763 | 3834 | 3905 | 3976 | 4047 | 4118 | 4189 | 4260 | **71** |
| **72** | 3672 | 3744 | 3816 | 3888 | 3960 | 4032 | 4104 | 4176 | 4248 | 4320 | **72** |
| **73** | 3723 | 3796 | 3869 | 3942 | 4015 | 4088 | 4161 | 4234 | 4307 | 4380 | **73** |
| **74** | 3774 | 3848 | 3922 | 3996 | 4070 | 4144 | 4218 | 4292 | 4366 | 4440 | **74** |
| **75** | 3825 | 3900 | 3975 | 4050 | 4125 | 4200 | 4275 | 4350 | 4425 | 4500 | **75** |
| **76** | 3876 | 3952 | 4028 | 4104 | 4180 | 4256 | 4332 | 4408 | 4484 | 4560 | **76** |
| **77** | 3927 | 4004 | 4081 | 4158 | 4235 | 4312 | 4389 | 4466 | 4543 | 4620 | **77** |
| **78** | 3978 | 4056 | 4134 | 4212 | 4290 | 4368 | 4446 | 4524 | 4602 | 4680 | **78** |
| **79** | 4029 | 4108 | 4187 | 4266 | 4345 | 4424 | 4503 | 4582 | 4661 | 4740 | **79** |
| **80** | 4080 | 4160 | 4240 | 4320 | 4400 | 4480 | 4560 | 4640 | 4720 | 4800 | **80** |
| **81** | 4131 | 4212 | 4293 | 4374 | 4455 | 4536 | 4617 | 4698 | 4779 | 4860 | **81** |
| **82** | 4182 | 4264 | 4346 | 4428 | 4510 | 4592 | 4674 | 4756 | 4838 | 4920 | **82** |
| **83** | 4233 | 4316 | 4399 | 4482 | 4565 | 4648 | 4731 | 4814 | 4897 | 4980 | **83** |
| **84** | 4284 | 4368 | 4452 | 4536 | 4620 | 4704 | 4788 | 4872 | 4956 | 5040 | **84** |
| **85** | 4335 | 4420 | 4505 | 4590 | 4675 | 4760 | 4845 | 4930 | 5015 | 5100 | **85** |
| **86** | 4386 | 4472 | 4558 | 4644 | 4730 | 4816 | 4902 | 4988 | 5074 | 5160 | **86** |
| **87** | 4437 | 4524 | 4611 | 4698 | 4785 | 4872 | 4959 | 5046 | 5133 | 5220 | **87** |
| **88** | 4488 | 4576 | 4664 | 4752 | 4840 | 4928 | 5016 | 5104 | 5192 | 5280 | **88** |
| **89** | 4539 | 4628 | 4717 | 4806 | 4895 | 4984 | 5073 | 5162 | 5251 | 5340 | **89** |
| **90** | 4590 | 4680 | 4770 | 4860 | 4950 | 5040 | 5130 | 5220 | 5310 | 5400 | **90** |
| **91** | 4641 | 4732 | 4823 | 4914 | 5005 | 5096 | 5187 | 5278 | 5369 | 5460 | **91** |
| **92** | 4692 | 4784 | 4876 | 4968 | 5060 | 5152 | 5244 | 5336 | 5428 | 5520 | **92** |
| **93** | 4743 | 4836 | 4929 | 5022 | 5115 | 5208 | 5301 | 5394 | 5487 | 5580 | **93** |
| **94** | 4794 | 4888 | 4982 | 5076 | 5170 | 5264 | 5358 | 5452 | 5546 | 5640 | **94** |
| **95** | 4845 | 4940 | 5035 | 5130 | 5225 | 5320 | 5415 | 5510 | 5605 | 5700 | **95** |
| **96** | 4896 | 4992 | 5088 | 5184 | 5280 | 5376 | 5472 | 5568 | 5664 | 5760 | **96** |
| **97** | 4947 | 5044 | 5141 | 5238 | 5335 | 5432 | 5529 | 5626 | 5723 | 5820 | **97** |
| **98** | 4998 | 5096 | 5194 | 5292 | 5390 | 5488 | 5586 | 5684 | 5782 | 5880 | **98** |
| **99** | 5049 | 5148 | 5247 | 5346 | 5445 | 5544 | 5643 | 5742 | 5841 | 5940 | **99** |
| **100** | 5100 | 5200 | 5300 | 5400 | 5500 | 5600 | 5700 | 5800 | 5900 | 6000 | **100** |
| | **51** | **52** | **53** | **54** | **55** | **56** | **57** | **58** | **59** | **60** | |

| | 61 | 62 | 63 | 64 | 65 | 66 | 67 | 68 | 69 | 70 | |
|---|---|---|---|---|---|---|---|---|---|---|---|
| 1 | 61 | 62 | 63 | 64 | 65 | 66 | 67 | 68 | 69 | 70 | 1 |
| 2 | 122 | 124 | 126 | 128 | 130 | 132 | 134 | 136 | 138 | 140 | 2 |
| 3 | 183 | 186 | 189 | 192 | 195 | 198 | 201 | 204 | 207 | 210 | 3 |
| 4 | 244 | 248 | 252 | 256 | 260 | 264 | 268 | 272 | 276 | 280 | 4 |
| 5 | 305 | 310 | 315 | 320 | 325 | 330 | 335 | 340 | 345 | 350 | 5 |
| 6 | 366 | 372 | 378 | 384 | 390 | 396 | 402 | 408 | 414 | 420 | 6 |
| 7 | 427 | 434 | 441 | 448 | 455 | 462 | 469 | 476 | 483 | 490 | 7 |
| 8 | 488 | 496 | 504 | 512 | 520 | 528 | 536 | 544 | 552 | 560 | 8 |
| 9 | 549 | 558 | 567 | 576 | 585 | 594 | 603 | 612 | 621 | 630 | 9 |
| 10 | 610 | 620 | 630 | 640 | 650 | 660 | 670 | 680 | 690 | 700 | 10 |
| 11 | 671 | 682 | 693 | 704 | 715 | 726 | 737 | 748 | 759 | 770 | 11 |
| 12 | 732 | 744 | 756 | 768 | 780 | 792 | 804 | 816 | 828 | 840 | 12 |
| 13 | 793 | 806 | 819 | 832 | 845 | 858 | 871 | 884 | 897 | 910 | 13 |
| 14 | 854 | 868 | 882 | 896 | 910 | 924 | 938 | 952 | 966 | 980 | 14 |
| 15 | 915 | 930 | 945 | 960 | 975 | 990 | 1005 | 1020 | 1035 | 1050 | 15 |
| 16 | 976 | 992 | 1008 | 1024 | 1040 | 1056 | 1072 | 1088 | 1104 | 1120 | 16 |
| 17 | 1037 | 1054 | 1071 | 1088 | 1105 | 1122 | 1139 | 1156 | 1173 | 1190 | 17 |
| 18 | 1098 | 1116 | 1134 | 1152 | 1170 | 1188 | 1206 | 1224 | 1242 | 1260 | 18 |
| 19 | 1159 | 1178 | 1197 | 1216 | 1235 | 1254 | 1273 | 1292 | 1311 | 1330 | 19 |
| 20 | 1220 | 1240 | 1260 | 1280 | 1300 | 1320 | 1340 | 1360 | 1380 | 1400 | 20 |
| 21 | 1281 | 1302 | 1323 | 1344 | 1365 | 1386 | 1407 | 1428 | 1449 | 1470 | 21 |
| 22 | 1342 | 1364 | 1386 | 1408 | 1430 | 1452 | 1474 | 1496 | 1518 | 1540 | 22 |
| 23 | 1403 | 1426 | 1449 | 1472 | 1495 | 1518 | 1541 | 1564 | 1587 | 1610 | 23 |
| 24 | 1464 | 1488 | 1512 | 1536 | 1560 | 1584 | 1608 | 1632 | 1656 | 1680 | 24 |
| 25 | 1525 | 1550 | 1575 | 1600 | 1625 | 1650 | 1675 | 1700 | 1725 | 1750 | 25 |
| 26 | 1586 | 1612 | 1638 | 1664 | 1690 | 1716 | 1742 | 1768 | 1794 | 1820 | 26 |
| 27 | 1647 | 1674 | 1701 | 1728 | 1755 | 1782 | 1809 | 1836 | 1863 | 1890 | 27 |
| 28 | 1708 | 1736 | 1764 | 1792 | 1820 | 1848 | 1876 | 1904 | 1932 | 1960 | 28 |
| 29 | 1769 | 1798 | 1827 | 1856 | 1885 | 1914 | 1943 | 1972 | 2001 | 2030 | 29 |
| 30 | 1830 | 1860 | 1890 | 1920 | 1950 | 1980 | 2010 | 2040 | 2070 | 2100 | 30 |
| 31 | 1891 | 1922 | 1953 | 1984 | 2015 | 2046 | 2077 | 2108 | 2139 | 2170 | 31 |
| 32 | 1952 | 1984 | 2016 | 2048 | 2080 | 2112 | 2144 | 2176 | 2208 | 2240 | 32 |
| 33 | 2013 | 2046 | 2079 | 2112 | 2145 | 2178 | 2211 | 2244 | 2277 | 2310 | 33 |
| 34 | 2074 | 2108 | 2142 | 2176 | 2210 | 2244 | 2278 | 2312 | 2346 | 2380 | 34 |
| 35 | 2135 | 2170 | 2205 | 2240 | 2275 | 2310 | 2345 | 2380 | 2415 | 2450 | 35 |
| 36 | 2196 | 2232 | 2268 | 2304 | 2340 | 2376 | 2412 | 2448 | 2484 | 2520 | 36 |
| 37 | 2257 | 2294 | 2331 | 2368 | 2405 | 2442 | 2479 | 2516 | 2553 | 2590 | 37 |
| 38 | 2318 | 2356 | 2394 | 2432 | 2470 | 2508 | 2546 | 2584 | 2622 | 2660 | 38 |
| 39 | 2379 | 2418 | 2457 | 2496 | 2535 | 2574 | 2613 | 2652 | 2691 | 2730 | 39 |
| 40 | 2440 | 2480 | 2520 | 2560 | 2600 | 2640 | 2680 | 2720 | 2760 | 2800 | 40 |
| 41 | 2501 | 2542 | 2583 | 2624 | 2665 | 2706 | 2747 | 2788 | 2829 | 2870 | 41 |
| 42 | 2562 | 2604 | 2646 | 2688 | 2730 | 2772 | 2814 | 2856 | 2898 | 2940 | 42 |
| 43 | 2623 | 2666 | 2709 | 2752 | 2795 | 2838 | 2881 | 2924 | 2967 | 3010 | 43 |
| 44 | 2684 | 2728 | 2772 | 2816 | 2860 | 2904 | 2948 | 2992 | 3036 | 3080 | 44 |
| 45 | 2745 | 2790 | 2835 | 2880 | 2925 | 2970 | 3015 | 3060 | 3105 | 3150 | 45 |
| 46 | 2806 | 2852 | 2898 | 2944 | 2990 | 3036 | 3082 | 3128 | 3174 | 3220 | 46 |
| 47 | 2867 | 2914 | 2961 | 3008 | 3055 | 3102 | 3149 | 3196 | 3243 | 3290 | 47 |
| 48 | 2928 | 2976 | 3024 | 3072 | 3120 | 3168 | 3216 | 3264 | 3312 | 3360 | 48 |
| 49 | 2989 | 3038 | 3087 | 3136 | 3185 | 3234 | 3283 | 3332 | 3381 | 3430 | 49 |
| 50 | 3050 | 3100 | 3150 | 3200 | 3250 | 3300 | 3350 | 3400 | 3450 | 3500 | 50 |
| | 61 | 62 | 63 | 64 | 65 | 66 | 67 | 68 | 69 | 70 | |

| | 61 | 62 | 63 | 64 | 65 | 66 | 67 | 68 | 69 | 70 | |
|---|---|---|---|---|---|---|---|---|---|---|---|
| 51 | 3111 | 3162 | 3213 | 3264 | 3315 | 3366 | 3417 | 3468 | 3519 | 3570 | 51 |
| 52 | 3172 | 3224 | 3276 | 3328 | 3380 | 3432 | 3484 | 3536 | 3588 | 3640 | 52 |
| 53 | 3233 | 3286 | 3339 | 3392 | 3445 | 3498 | 3551 | 3604 | 3657 | 3710 | 53 |
| 54 | 3294 | 3348 | 3402 | 3456 | 3510 | 3564 | 3618 | 3672 | 3726 | 3780 | 54 |
| 55 | 3355 | 3410 | 3465 | 3520 | 3575 | 3630 | 3685 | 3740 | 3795 | 3850 | 55 |
| 56 | 3416 | 3472 | 3528 | 3584 | 3640 | 3696 | 3752 | 3808 | 3864 | 3920 | 56 |
| 57 | 3477 | 3534 | 3591 | 3648 | 3705 | 3762 | 3819 | 3876 | 3933 | 3990 | 57 |
| 58 | 3538 | 3596 | 3654 | 3712 | 3770 | 3828 | 3886 | 3944 | 4002 | 4060 | 58 |
| 59 | 3599 | 3658 | 3717 | 3776 | 3835 | 3894 | 3953 | 4012 | 4071 | 4130 | 59 |
| 60 | 3660 | 3720 | 3780 | 3840 | 3900 | 3960 | 4020 | 4080 | 4140 | 4200 | 60 |
| 61 | 3721 | 3782 | 3843 | 3904 | 3965 | 4026 | 4087 | 4148 | 4209 | 4270 | 61 |
| 62 | 3782 | 3844 | 3906 | 3968 | 4030 | 4092 | 4154 | 4216 | 4278 | 4340 | 62 |
| 63 | 3843 | 3906 | 3969 | 4032 | 4095 | 4158 | 4221 | 4284 | 4347 | 4410 | 63 |
| 64 | 3904 | 3968 | 4032 | 4096 | 4160 | 4224 | 4288 | 4352 | 4416 | 4480 | 64 |
| 65 | 3965 | 4030 | 4095 | 4160 | 4225 | 4290 | 4355 | 4420 | 4485 | 4550 | 65 |
| 66 | 4026 | 4092 | 4158 | 4224 | 4290 | 4356 | 4422 | 4488 | 4554 | 4620 | 66 |
| 67 | 4087 | 4154 | 4221 | 4288 | 4355 | 4422 | 4489 | 4556 | 4623 | 4690 | 67 |
| 68 | 4148 | 4216 | 4284 | 4352 | 4420 | 4488 | 4556 | 4624 | 4692 | 4760 | 68 |
| 69 | 4209 | 4278 | 4347 | 4416 | 4485 | 4554 | 4623 | 4692 | 4761 | 4830 | 69 |
| 70 | 4270 | 4340 | 4410 | 4480 | 4550 | 4620 | 4690 | 4760 | 4830 | 4900 | 70 |
| 71 | 4331 | 4402 | 4473 | 4544 | 4615 | 4686 | 4757 | 4828 | 4899 | 4970 | 71 |
| 72 | 4392 | 4464 | 4536 | 4608 | 4680 | 4752 | 4824 | 4896 | 4968 | 5040 | 72 |
| 73 | 4453 | 4526 | 4599 | 4672 | 4745 | 4818 | 4891 | 4964 | 5037 | 5110 | 73 |
| 74 | 4514 | 4588 | 4662 | 4736 | 4810 | 4884 | 4958 | 5032 | 5106 | 5180 | 74 |
| 75 | 4575 | 4650 | 4725 | 4800 | 4875 | 4950 | 5025 | 5100 | 5175 | 5250 | 75 |
| 76 | 4636 | 4712 | 4788 | 4864 | 4940 | 5016 | 5092 | 5168 | 5244 | 5320 | 76 |
| 77 | 4697 | 4774 | 4851 | 4928 | 5005 | 5082 | 5159 | 5236 | 5313 | 5390 | 77 |
| 78 | 4758 | 4836 | 4914 | 4992 | 5070 | 5148 | 5226 | 5304 | 5382 | 5460 | 78 |
| 79 | 4819 | 4898 | 4977 | 5056 | 5135 | 5214 | 5293 | 5372 | 5451 | 5530 | 79 |
| 80 | 4880 | 4960 | 5040 | 5120 | 5200 | 5280 | 5360 | 5440 | 5520 | 5600 | 80 |
| 81 | 4941 | 5022 | 5103 | 5184 | 5265 | 5346 | 5427 | 5508 | 5589 | 5670 | 81 |
| 82 | 5002 | 5084 | 5166 | 5248 | 5330 | 5412 | 5494 | 5576 | 5658 | 5740 | 82 |
| 83 | 5063 | 5146 | 5229 | 5312 | 5395 | 5478 | 5561 | 5644 | 5727 | 5810 | 83 |
| 84 | 5124 | 5208 | 5292 | 5376 | 5460 | 5544 | 5628 | 5712 | 5796 | 5880 | 84 |
| 85 | 5185 | 5270 | 5355 | 5440 | 5525 | 5610 | 5695 | 5780 | 5865 | 5950 | 85 |
| 86 | 5246 | 5332 | 5418 | 5504 | 5590 | 5676 | 5762 | 5848 | 5934 | 6020 | 86 |
| 87 | 5307 | 5394 | 5481 | 5568 | 5655 | 5742 | 5829 | 5916 | 6003 | 6090 | 87 |
| 88 | 5368 | 5456 | 5544 | 5632 | 5720 | 5808 | 5896 | 5984 | 6072 | 6160 | 88 |
| 89 | 5429 | 5518 | 5607 | 5696 | 5785 | 5874 | 5963 | 6052 | 6141 | 6230 | 89 |
| 90 | 5490 | 5580 | 5670 | 5760 | 5850 | 5940 | 6030 | 6120 | 6210 | 6300 | 90 |
| 91 | 5551 | 5642 | 5733 | 5824 | 5915 | 6006 | 6097 | 6188 | 6279 | 6370 | 91 |
| 92 | 5612 | 5704 | 5796 | 5888 | 5980 | 6072 | 6164 | 6256 | 6348 | 6440 | 92 |
| 93 | 5673 | 5766 | 5859 | 5952 | 6045 | 6138 | 6231 | 6324 | 6417 | 6510 | 93 |
| 94 | 5734 | 5828 | 5922 | 6016 | 6110 | 6204 | 6298 | 6392 | 6486 | 6580 | 94 |
| 95 | 5795 | 5890 | 5985 | 6080 | 6175 | 6270 | 6365 | 6460 | 6555 | 6650 | 95 |
| 96 | 5856 | 5952 | 6048 | 6144 | 6240 | 6336 | 6432 | 6528 | 6624 | 6720 | 96 |
| 97 | 5917 | 6014 | 6111 | 6208 | 6305 | 6402 | 6499 | 6596 | 6693 | 6790 | 97 |
| 98 | 5978 | 6076 | 6174 | 6272 | 6370 | 6468 | 6566 | 6664 | 6762 | 6860 | 98 |
| 99 | 6039 | 6138 | 6237 | 6336 | 6435 | 6534 | 6633 | 6732 | 6831 | 6930 | 99 |
| 100 | 6100 | 6200 | 6300 | 6400 | 6500 | 6600 | 6700 | 6800 | 6900 | 7000 | 100 |
| | 61 | 62 | 63 | 64 | 65 | 66 | 67 | 68 | 69 | 70 | |

| | 71 | 72 | 73 | 74 | 75 | 76 | 77 | 78 | 79 | 80 | |
|---|---|---|---|---|---|---|---|---|---|---|---|
| 1 | 71 | 72 | 73 | 74 | 75 | 76 | 77 | 78 | 79 | 80 | 1 |
| 2 | 142 | 144 | 146 | 148 | 150 | 152 | 154 | 156 | 158 | 160 | 2 |
| 3 | 213 | 216 | 219 | 222 | 225 | 228 | 231 | 234 | 237 | 240 | 3 |
| 4 | 284 | 288 | 292 | 296 | 300 | 304 | 308 | 312 | 316 | 320 | 4 |
| 5 | 355 | 360 | 365 | 370 | 375 | 380 | 385 | 390 | 395 | 400 | 5 |
| 6 | 426 | 432 | 438 | 444 | 450 | 456 | 462 | 468 | 474 | 480 | 6 |
| 7 | 497 | 504 | 511 | 518 | 525 | 532 | 539 | 546 | 553 | 560 | 7 |
| 8 | 568 | 576 | 584 | 592 | 600 | 608 | 616 | 624 | 632 | 640 | 8 |
| 9 | 639 | 648 | 657 | 666 | 675 | 684 | 693 | 702 | 711 | 720 | 9 |
| 10 | 710 | 720 | 730 | 740 | 750 | 760 | 770 | 780 | 790 | 800 | 10 |
| 11 | 781 | 792 | 803 | 814 | 825 | 836 | 847 | 858 | 869 | 880 | 11 |
| 12 | 852 | 864 | 876 | 888 | 900 | 912 | 924 | 936 | 948 | 960 | 12 |
| 13 | 923 | 936 | 949 | 962 | 975 | 988 | 1001 | 1014 | 1027 | 1040 | 13 |
| 14 | 994 | 1008 | 1022 | 1036 | 1050 | 1064 | 1078 | 1092 | 1106 | 1120 | 14 |
| 15 | 1065 | 1080 | 1095 | 1110 | 1125 | 1140 | 1155 | 1170 | 1185 | 1200 | 15 |
| 16 | 1136 | 1152 | 1168 | 1184 | 1200 | 1216 | 1232 | 1248 | 1264 | 1280 | 16 |
| 17 | 1207 | 1224 | 1241 | 1258 | 1275 | 1292 | 1309 | 1326 | 1343 | 1360 | 17 |
| 18 | 1278 | 1296 | 1314 | 1332 | 1350 | 1368 | 1386 | 1404 | 1422 | 1440 | 18 |
| 19 | 1349 | 1368 | 1387 | 1406 | 1425 | 1444 | 1463 | 1482 | 1501 | 1520 | 19 |
| 20 | 1420 | 1440 | 1460 | 1480 | 1500 | 1520 | 1540 | 1560 | 1580 | 1600 | 20 |
| 21 | 1491 | 1512 | 1533 | 1554 | 1575 | 1596 | 1617 | 1638 | 1659 | 1680 | 21 |
| 22 | 1562 | 1584 | 1606 | 1628 | 1650 | 1672 | 1694 | 1716 | 1738 | 1760 | 22 |
| 23 | 1633 | 1656 | 1679 | 1702 | 1725 | 1748 | 1771 | 1794 | 1817 | 1840 | 23 |
| 24 | 1704 | 1728 | 1752 | 1776 | 1800 | 1824 | 1848 | 1872 | 1896 | 1920 | 24 |
| 25 | 1775 | 1800 | 1825 | 1850 | 1875 | 1900 | 1925 | 1950 | 1975 | 2000 | 25 |
| 26 | 1846 | 1872 | 1898 | 1924 | 1950 | 1976 | 2002 | 2028 | 2054 | 2080 | 26 |
| 27 | 1917 | 1944 | 1971 | 1998 | 2025 | 2052 | 2079 | 2106 | 2133 | 2160 | 27 |
| 28 | 1988 | 2016 | 2044 | 2072 | 2100 | 2128 | 2156 | 2184 | 2212 | 2240 | 28 |
| 29 | 2059 | 2088 | 2117 | 2146 | 2175 | 2204 | 2233 | 2262 | 2291 | 2320 | 29 |
| 30 | 2130 | 2160 | 2190 | 2220 | 2250 | 2280 | 2310 | 2340 | 2370 | 2400 | 30 |
| 31 | 2201 | 2232 | 2263 | 2294 | 2325 | 2356 | 2387 | 2418 | 2449 | 2480 | 31 |
| 32 | 2272 | 2304 | 2336 | 2368 | 2400 | 2432 | 2464 | 2496 | 2528 | 2560 | 32 |
| 33 | 2343 | 2376 | 2409 | 2442 | 2475 | 2508 | 2541 | 2574 | 2607 | 2640 | 33 |
| 34 | 2414 | 2448 | 2482 | 2516 | 2550 | 2584 | 2618 | 2652 | 2686 | 2720 | 34 |
| 35 | 2485 | 2520 | 2555 | 2590 | 2625 | 2660 | 2695 | 2730 | 2765 | 2800 | 35 |
| 36 | 2556 | 2592 | 2628 | 2664 | 2700 | 2736 | 2772 | 2808 | 2844 | 2880 | 36 |
| 37 | 2627 | 2664 | 2701 | 2738 | 2775 | 2812 | 2849 | 2886 | 2923 | 2960 | 37 |
| 38 | 2698 | 2736 | 2774 | 2812 | 2850 | 2888 | 2926 | 2964 | 3002 | 3040 | 38 |
| 39 | 2769 | 2808 | 2847 | 2886 | 2925 | 2964 | 3003 | 3042 | 3081 | 3120 | 39 |
| 40 | 2840 | 2880 | 2920 | 2960 | 3000 | 3040 | 3080 | 3120 | 3160 | 3200 | 40 |
| 41 | 2911 | 2952 | 2993 | 3034 | 3075 | 3116 | 3157 | 3198 | 3239 | 3280 | 41 |
| 42 | 2982 | 3024 | 3066 | 3108 | 3150 | 3192 | 3234 | 3276 | 3318 | 3360 | 42 |
| 43 | 3053 | 3096 | 3139 | 3182 | 3225 | 3268 | 3311 | 3354 | 3397 | 3440 | 43 |
| 44 | 3124 | 3168 | 3212 | 3256 | 3300 | 3344 | 3388 | 3432 | 3476 | 3520 | 44 |
| 45 | 3195 | 3240 | 3285 | 3330 | 3375 | 3420 | 3465 | 3510 | 3555 | 3600 | 45 |
| 46 | 3266 | 3312 | 3358 | 3404 | 3450 | 3496 | 3542 | 3588 | 3634 | 3680 | 46 |
| 47 | 3337 | 3384 | 3431 | 3478 | 3525 | 3572 | 3619 | 3666 | 3713 | 3760 | 47 |
| 48 | 3408 | 3456 | 3504 | 3552 | 3600 | 3648 | 3696 | 3744 | 3792 | 3840 | 48 |
| 49 | 3479 | 3528 | 3577 | 3626 | 3675 | 3724 | 3773 | 3822 | 3871 | 3920 | 49 |
| 50 | 3550 | 3600 | 3650 | 3700 | 3750 | 3800 | 3850 | 3900 | 3950 | 4000 | 50 |
| | 71 | 72 | 73 | 74 | 75 | 76 | 77 | 78 | 79 | 80 | |

| | 71 | 72 | 73 | 74 | 75 | 76 | 77 | 78 | 79 | 80 | |
|---|---|---|---|---|---|---|---|---|---|---|---|
| **51** | 3621 | 3672 | 3723 | 3774 | 3825 | 3876 | 3927 | 3978 | 4029 | 4080 | **51** |
| **52** | 3692 | 3744 | 3796 | 3848 | 3900 | 3952 | 4004 | 4056 | 4108 | 4160 | **52** |
| **53** | 3763 | 3816 | 3869 | 3922 | 3975 | 4028 | 4081 | 4134 | 4187 | 4240 | **53** |
| **54** | 3834 | 3888 | 3942 | 3996 | 4050 | 4104 | 4158 | 4212 | 4266 | 4320 | **54** |
| **55** | 3905 | 3960 | 4015 | 4070 | 4125 | 4180 | 4235 | 4290 | 4345 | 4400 | **55** |
| **56** | 3976 | 4032 | 4088 | 4144 | 4200 | 4256 | 4312 | 4368 | 4424 | 4480 | **56** |
| **57** | 4047 | 4104 | 4161 | 4218 | 4275 | 4332 | 4389 | 4446 | 4503 | 4560 | **57** |
| **58** | 4118 | 4176 | 4234 | 4292 | 4350 | 4408 | 4466 | 4524 | 4582 | 4640 | **58** |
| **59** | 4189 | 4248 | 4307 | 4366 | 4425 | 4484 | 4543 | 4602 | 4661 | 4720 | **59** |
| **60** | 4260 | 4320 | 4380 | 4440 | 4500 | 4560 | 4620 | 4680 | 4740 | 4800 | **60** |
| **61** | 4331 | 4392 | 4453 | 4514 | 4575 | 4636 | 4697 | 4758 | 4819 | 4880 | **61** |
| **62** | 4402 | 4464 | 4526 | 4588 | 4650 | 4712 | 4774 | 4836 | 4898 | 4960 | **62** |
| **63** | 4473 | 4536 | 4599 | 4662 | 4725 | 4788 | 4851 | 4914 | 4977 | 5040 | **63** |
| **64** | 4544 | 4608 | 4672 | 4736 | 4800 | 4864 | 4928 | 4992 | 5056 | 5120 | **64** |
| **65** | 4615 | 4680 | 4745 | 4810 | 4875 | 4940 | 5005 | 5070 | 5135 | 5200 | **65** |
| **66** | 4686 | 4752 | 4818 | 4884 | 4950 | 5016 | 5082 | 5148 | 5214 | 5280 | **66** |
| **67** | 4757 | 4824 | 4891 | 4958 | 5025 | 5092 | 5159 | 5226 | 5293 | 5360 | **67** |
| **68** | 4828 | 4896 | 4964 | 5032 | 5100 | 5168 | 5236 | 5304 | 5372 | 5440 | **68** |
| **69** | 4899 | 4968 | 5037 | 5106 | 5175 | 5244 | 5313 | 5382 | 5451 | 5520 | **69** |
| **70** | 4970 | 5040 | 5110 | 5180 | 5250 | 5320 | 5390 | 5460 | 5530 | 5600 | **70** |
| **71** | 5041 | 5112 | 5183 | 5254 | 5325 | 5396 | 5467 | 5538 | 5609 | 5680 | **71** |
| **72** | 5112 | 5184 | 5256 | 5328 | 5400 | 5472 | 5544 | 5616 | 5688 | 5760 | **72** |
| **73** | 5183 | 5256 | 5329 | 5402 | 5475 | 5548 | 5621 | 5694 | 5767 | 5840 | **73** |
| **74** | 5254 | 5328 | 5402 | 5476 | 5550 | 5624 | 5698 | 5772 | 5846 | 5920 | **74** |
| **75** | 5325 | 5400 | 5475 | 5550 | 5625 | 5700 | 5775 | 5850 | 5925 | 6000 | **75** |
| **76** | 5396 | 5472 | 5548 | 5624 | 5700 | 5776 | 5852 | 5928 | 6004 | 6080 | **76** |
| **77** | 5467 | 5544 | 5621 | 5698 | 5775 | 5852 | 5929 | 6006 | 6083 | 6160 | **77** |
| **78** | 5538 | 5616 | 5694 | 5772 | 5850 | 5928 | 6006 | 6084 | 6162 | 6240 | **78** |
| **79** | 5609 | 5688 | 5767 | 5846 | 5925 | 6004 | 6083 | 6162 | 6241 | 6320 | **79** |
| **80** | 5680 | 5760 | 5840 | 5920 | 6000 | 6080 | 6160 | 6240 | 6320 | 6400 | **80** |
| **81** | 5751 | 5832 | 5913 | 5994 | 6075 | 6156 | 6237 | 6318 | 6399 | 6480 | **81** |
| **82** | 5822 | 5904 | 5986 | 6068 | 6150 | 6232 | 6314 | 6396 | 6478 | 6560 | **82** |
| **83** | 5893 | 5976 | 6059 | 6142 | 6225 | 6308 | 6391 | 6474 | 6557 | 6640 | **83** |
| **84** | 5964 | 6048 | 6132 | 6216 | 6300 | 6384 | 6468 | 6552 | 6636 | 6720 | **84** |
| **85** | 6035 | 6120 | 6205 | 6290 | 6375 | 6460 | 6545 | 6630 | 6715 | 6800 | **85** |
| **86** | 6106 | 6192 | 6278 | 6364 | 6450 | 6536 | 6622 | 6708 | 6794 | 6880 | **86** |
| **87** | 6177 | 6264 | 6351 | 6438 | 6525 | 6612 | 6699 | 6786 | 6873 | 6960 | **87** |
| **88** | 6248 | 6336 | 6424 | 6512 | 6600 | 6688 | 6776 | 6864 | 6952 | 7040 | **88** |
| **89** | 6319 | 6408 | 6497 | 6586 | 6675 | 6764 | 6853 | 6942 | 7031 | 7120 | **89** |
| **90** | 6390 | 6480 | 6570 | 6660 | 6750 | 6840 | 6930 | 7020 | 7110 | 7200 | **90** |
| **91** | 6461 | 6552 | 6643 | 6734 | 6825 | 6916 | 7007 | 7098 | 7189 | 7280 | **91** |
| **92** | 6532 | 6624 | 6716 | 6808 | 6900 | 6992 | 7084 | 7176 | 7268 | 7360 | **92** |
| **93** | 6603 | 6696 | 6789 | 6882 | 6975 | 7068 | 7161 | 7254 | 7347 | 7440 | **93** |
| **94** | 6674 | 6768 | 6862 | 6956 | 7050 | 7144 | 7238 | 7332 | 7426 | 7520 | **94** |
| **95** | 6745 | 6840 | 6935 | 7030 | 7125 | 7220 | 7315 | 7410 | 7505 | 7600 | **95** |
| **96** | 6816 | 6912 | 7008 | 7104 | 7200 | 7296 | 7392 | 7488 | 7584 | 7680 | **96** |
| **97** | 6887 | 6984 | 7081 | 7178 | 7275 | 7372 | 7469 | 7566 | 7663 | 7760 | **97** |
| **98** | 6958 | 7056 | 7154 | 7252 | 7350 | 7448 | 7546 | 7644 | 7742 | 7840 | **98** |
| **99** | 7029 | 7128 | 7227 | 7326 | 7425 | 7524 | 7623 | 7722 | 7821 | 7920 | **99** |
| **100** | 7100 | 7200 | 7300 | 7400 | 7500 | 7600 | 7700 | 7800 | 7900 | 8000 | **100** |
| | 71 | 72 | 73 | 74 | 75 | 76 | 77 | 78 | 79 | 80 | |

| | 81 | 82 | 83 | 84 | 85 | 86 | 87 | 88 | 89 | 90 | |
|---|---|---|---|---|---|---|---|---|---|---|---|
| 1 | 81 | 82 | 83 | 84 | 85 | 86 | 87 | 88 | 89 | 90 | 1 |
| 2 | 162 | 164 | 166 | 168 | 170 | 172 | 174 | 176 | 178 | 180 | 2 |
| 3 | 243 | 246 | 249 | 252 | 255 | 258 | 261 | 264 | 267 | 270 | 3 |
| 4 | 324 | 328 | 332 | 336 | 340 | 344 | 348 | 352 | 356 | 360 | 4 |
| 5 | 405 | 410 | 415 | 420 | 425 | 430 | 435 | 440 | 445 | 450 | 5 |
| 6 | 486 | 492 | 498 | 504 | 510 | 516 | 522 | 528 | 534 | 540 | 6 |
| 7 | 567 | 574 | 581 | 588 | 595 | 602 | 609 | 616 | 623 | 630 | 7 |
| 8 | 648 | 656 | 664 | 672 | 680 | 688 | 696 | 704 | 712 | 720 | 8 |
| 9 | 729 | 738 | 747 | 756 | 765 | 774 | 783 | 792 | 801 | 810 | 9 |
| 10 | 810 | 820 | 830 | 840 | 850 | 860 | 870 | 880 | 890 | 900 | 10 |
| 11 | 891 | 902 | 913 | 924 | 935 | 946 | 957 | 968 | 979 | 990 | 11 |
| 12 | 972 | 984 | 996 | 1008 | 1020 | 1032 | 1044 | 1056 | 1068 | 1080 | 12 |
| 13 | 1053 | 1066 | 1079 | 1092 | 1105 | 1118 | 1131 | 1144 | 1157 | 1170 | 13 |
| 14 | 1134 | 1148 | 1162 | 1176 | 1190 | 1204 | 1218 | 1232 | 1246 | 1260 | 14 |
| 15 | 1215 | 1230 | 1245 | 1260 | 1275 | 1290 | 1305 | 1320 | 1335 | 1350 | 15 |
| 16 | 1296 | 1312 | 1328 | 1344 | 1360 | 1376 | 1392 | 1408 | 1424 | 1440 | 16 |
| 17 | 1377 | 1394 | 1411 | 1428 | 1445 | 1462 | 1479 | 1496 | 1513 | 1530 | 17 |
| 18 | 1458 | 1476 | 1494 | 1512 | 1530 | 1548 | 1566 | 1584 | 1602 | 1620 | 18 |
| 19 | 1539 | 1558 | 1577 | 1596 | 1615 | 1634 | 1653 | 1672 | 1691 | 1710 | 19 |
| 20 | 1620 | 1640 | 1660 | 1680 | 1700 | 1720 | 1740 | 1760 | 1780 | 1800 | 20 |
| 21 | 1701 | 1722 | 1743 | 1764 | 1785 | 1806 | 1827 | 1848 | 1869 | 1890 | 21 |
| 22 | 1782 | 1804 | 1826 | 1848 | 1870 | 1892 | 1914 | 1936 | 1958 | 1980 | 22 |
| 23 | 1863 | 1886 | 1909 | 1932 | 1955 | 1978 | 2001 | 2024 | 2047 | 2070 | 23 |
| 24 | 1944 | 1968 | 1992 | 2016 | 2040 | 2064 | 2088 | 2112 | 2136 | 2160 | 24 |
| 25 | 2025 | 2050 | 2075 | 2100 | 2125 | 2150 | 2175 | 2200 | 2225 | 2250 | 25 |
| 26 | 2106 | 2132 | 2158 | 2184 | 2210 | 2236 | 2262 | 2288 | 2314 | 2340 | 26 |
| 27 | 2187 | 2214 | 2241 | 2268 | 2295 | 2322 | 2349 | 2376 | 2403 | 2430 | 27 |
| 28 | 2268 | 2296 | 2324 | 2352 | 2380 | 2408 | 2436 | 2464 | 2492 | 2520 | 28 |
| 29 | 2349 | 2378 | 2407 | 2436 | 2465 | 2494 | 2523 | 2552 | 2581 | 2610 | 29 |
| 30 | 2430 | 2460 | 2490 | 2520 | 2550 | 2580 | 2610 | 2640 | 2670 | 2700 | 30 |
| 31 | 2511 | 2542 | 2573 | 2604 | 2635 | 2666 | 2697 | 2728 | 2759 | 2790 | 31 |
| 32 | 2592 | 2624 | 2656 | 2688 | 2720 | 2752 | 2784 | 2816 | 2848 | 2880 | 32 |
| 33 | 2673 | 2706 | 2739 | 2772 | 2805 | 2838 | 2871 | 2904 | 2937 | 2970 | 33 |
| 34 | 2754 | 2788 | 2822 | 2856 | 2890 | 2924 | 2958 | 2992 | 3026 | 3060 | 34 |
| 35 | 2835 | 2870 | 2905 | 2940 | 2975 | 3010 | 3045 | 3080 | 3115 | 3150 | 35 |
| 36 | 2916 | 2952 | 2988 | 3024 | 3060 | 3096 | 3132 | 3168 | 3204 | 3240 | 36 |
| 37 | 2997 | 3034 | 3071 | 3108 | 3145 | 3182 | 3219 | 3256 | 3293 | 3330 | 37 |
| 38 | 3078 | 3116 | 3154 | 3192 | 3230 | 3268 | 3306 | 3344 | 3382 | 3420 | 38 |
| 39 | 3159 | 3198 | 3237 | 3276 | 3315 | 3354 | 3393 | 3432 | 3471 | 3510 | 39 |
| 40 | 3240 | 3280 | 3320 | 3360 | 3400 | 3440 | 3480 | 3520 | 3560 | 3600 | 40 |
| 41 | 3321 | 3362 | 3403 | 3444 | 3485 | 3526 | 3567 | 3608 | 3649 | 3690 | 41 |
| 42 | 3402 | 3444 | 3486 | 3528 | 3570 | 3612 | 3654 | 3696 | 3738 | 3780 | 42 |
| 43 | 3483 | 3526 | 3569 | 3612 | 3655 | 3698 | 3741 | 3784 | 3827 | 3870 | 43 |
| 44 | 3564 | 3608 | 3652 | 3696 | 3740 | 3784 | 3828 | 3872 | 3916 | 3960 | 44 |
| 45 | 3645 | 3690 | 3735 | 3780 | 3825 | 3870 | 3915 | 3960 | 4005 | 4050 | 45 |
| 46 | 3726 | 3772 | 3818 | 3864 | 3910 | 3956 | 4002 | 4048 | 4094 | 4140 | 46 |
| 47 | 3807 | 3854 | 3901 | 3948 | 3995 | 4042 | 4089 | 4136 | 4183 | 4230 | 47 |
| 48 | 3888 | 3936 | 3984 | 4032 | 4080 | 4128 | 4176 | 4224 | 4272 | 4320 | 48 |
| 49 | 3969 | 4018 | 4067 | 4116 | 4165 | 4214 | 4263 | 4312 | 4361 | 4410 | 49 |
| 50 | 4050 | 4100 | 4150 | 4200 | 4250 | 4300 | 4350 | 4400 | 4450 | 4500 | 50 |
| | 81 | 82 | 83 | 84 | 85 | 86 | 87 | 88 | 89 | 90 | |

| | 81 | 82 | 83 | 84 | 85 | 86 | 87 | 88 | 89 | 90 | |
|---|---|---|---|---|---|---|---|---|---|---|---|
| 51 | 4131 | 4182 | 4233 | 4284 | 4335 | 4386 | 4437 | 4488 | 4539 | 4590 | 51 |
| 52 | 4212 | 4264 | 4316 | 4368 | 4420 | 4472 | 4524 | 4576 | 4628 | 4680 | 52 |
| 53 | 4293 | 4346 | 4399 | 4452 | 4505 | 4558 | 4611 | 4664 | 4717 | 4770 | 53 |
| 54 | 4374 | 4428 | 4482 | 4536 | 4590 | 4644 | 4698 | 4752 | 4806 | 4860 | 54 |
| 55 | 4455 | 4510 | 4565 | 4620 | 4675 | 4730 | 4785 | 4840 | 4895 | 4950 | 55 |
| 56 | 4536 | 4592 | 4648 | 4704 | 4760 | 4816 | 4872 | 4928 | 4984 | 5040 | 56 |
| 57 | 4617 | 4674 | 4731 | 4788 | 4845 | 4902 | 4959 | 5016 | 5073 | 5130 | 57 |
| 58 | 4698 | 4756 | 4814 | 4872 | 4930 | 4988 | 5046 | 5104 | 5162 | 5220 | 58 |
| 59 | 4779 | 4838 | 4897 | 4956 | 5015 | 5074 | 5133 | 5192 | 5251 | 5310 | 59 |
| 60 | 4860 | 4920 | 4980 | 5040 | 5100 | 5160 | 5220 | 5280 | 5340 | 5400 | 60 |
| 61 | 4941 | 5002 | 5063 | 5124 | 5185 | 5246 | 5307 | 5368 | 5429 | 5490 | 61 |
| 62 | 5022 | 5084 | 5146 | 5208 | 5270 | 5332 | 5394 | 5456 | 5518 | 5580 | 62 |
| 63 | 5103 | 5166 | 5229 | 5292 | 5355 | 5418 | 5481 | 5544 | 5607 | 5670 | 63 |
| 64 | 5184 | 5248 | 5312 | 5376 | 5440 | 5504 | 5568 | 5632 | 5696 | 5760 | 64 |
| 65 | 5265 | 5330 | 5395 | 5460 | 5525 | 5590 | 5655 | 5720 | 5785 | 5850 | 65 |
| 66 | 5346 | 5412 | 5478 | 5544 | 5610 | 5676 | 5742 | 5808 | 5874 | 5940 | 66 |
| 67 | 5427 | 5494 | 5561 | 5628 | 5695 | 5762 | 5829 | 5896 | 5963 | 6030 | 67 |
| 68 | 5508 | 5576 | 5644 | 5712 | 5780 | 5848 | 5916 | 5984 | 6052 | 6120 | 68 |
| 69 | 5589 | 5658 | 5727 | 5796 | 5865 | 5934 | 6003 | 6072 | 6141 | 6210 | 89 |
| 70 | 5670 | 5740 | 5810 | 5880 | 5950 | 6020 | 6090 | 6160 | 6230 | 6300 | 70 |
| 71 | 5751 | 5822 | 5893 | 5964 | 6035 | 6106 | 6177 | 6248 | 6319 | 6390 | 71 |
| 72 | 5832 | 5904 | 5976 | 6048 | 6120 | 6192 | 6264 | 6336 | 6408 | 6480 | 72 |
| 73 | 5913 | 5986 | 6059 | 6132 | 6205 | 6278 | 6351 | 6424 | 6497 | 6570 | 73 |
| 74 | 5994 | 6068 | 6142 | 6216 | 6290 | 6364 | 6438 | 6512 | 6586 | 6660 | 74 |
| 75 | 6075 | 6150 | 6225 | 6300 | 6375 | 6450 | 6525 | 6600 | 6675 | 6750 | 75 |
| 76 | 6156 | 6232 | 6308 | 6384 | 6460 | 6536 | 6612 | 6688 | 6764 | 6840 | 76 |
| 77 | 6237 | 6314 | 6391 | 6468 | 6545 | 6622 | 6699 | 6776 | 6853 | 6930 | 77 |
| 78 | 6318 | 6396 | 6474 | 6552 | 6630 | 6708 | 6786 | 6864 | 6942 | 7020 | 78 |
| 79 | 6399 | 6478 | 6557 | 6636 | 6715 | 6794 | 6873 | 6952 | 7031 | 7110 | 79 |
| 80 | 6480 | 6560 | 6640 | 6720 | 6800 | 6880 | 6960 | 7040 | 7120 | 7200 | 80 |
| 81 | 6561 | 6642 | 6723 | 6804 | 6885 | 6966 | 7047 | 7128 | 7209 | 7290 | 81 |
| 82 | 6642 | 6724 | 6806 | 6888 | 6970 | 7052 | 7134 | 7216 | 7298 | 7380 | 82 |
| 83 | 6723 | 6806 | 6889 | 6972 | 7055 | 7138 | 7221 | 7304 | 7387 | 7470 | 83 |
| 84 | 6804 | 6888 | 6972 | 7056 | 7140 | 7224 | 7308 | 7392 | 7476 | 7560 | 84 |
| 85 | 6885 | 6970 | 7055 | 7140 | 7225 | 7310 | 7395 | 7480 | 7565 | 7650 | 85 |
| 86 | 6966 | 7052 | 7138 | 7224 | 7310 | 7396 | 7482 | 7568 | 7654 | 7740 | 86 |
| 87 | 7047 | 7134 | 7221 | 7308 | 7395 | 7482 | 7569 | 7656 | 7743 | 7830 | 87 |
| 88 | 7128 | 7216 | 7304 | 7392 | 7480 | 7568 | 7656 | 7744 | 7832 | 7920 | 88 |
| 89 | 7209 | 7298 | 7387 | 7476 | 7565 | 7654 | 7743 | 7832 | 7921 | 8010 | 89 |
| 90 | 7290 | 7380 | 7470 | 7560 | 7650 | 7740 | 7830 | 7920 | 8010 | 8100 | 90 |
| 91 | 7371 | 7462 | 7553 | 7644 | 7735 | 7826 | 7917 | 8008 | 8099 | 8190 | 91 |
| 92 | 7452 | 7544 | 7636 | 7728 | 7820 | 7912 | 8004 | 8096 | 8188 | 8280 | 92 |
| 93 | 7533 | 7626 | 7719 | 7812 | 7905 | 7998 | 8091 | 8184 | 8277 | 8370 | 93 |
| 94 | 7614 | 7708 | 7802 | 7896 | 7990 | 8084 | 8178 | 8272 | 8366 | 8460 | 94 |
| 95 | 7695 | 7790 | 7885 | 7980 | 8075 | 8170 | 8265 | 8360 | 8455 | 8550 | 95 |
| 96 | 7776 | 7872 | 7968 | 8064 | 8160 | 8256 | 8352 | 8448 | 8544 | 8640 | 96 |
| 97 | 7857 | 7954 | 8051 | 8148 | 8245 | 8342 | 8439 | 8536 | 8633 | 8730 | 97 |
| 98 | 7938 | 8036 | 8134 | 8232 | 8330 | 8428 | 8526 | 8624 | 8722 | 8820 | 98 |
| 99 | 8019 | 8118 | 8217 | 8316 | 8415 | 8514 | 8613 | 8712 | 8811 | 8910 | 99 |
| 100 | 8100 | 8200 | 8300 | 8400 | 8500 | 8600 | 8700 | 9800 | 8900 | 9000 | 100 |
| | 81 | 82 | 83 | 84 | 85 | 86 | 87 | 88 | 89 | 90 | |

| | 91 | 92 | 93 | 94 | 95 | 96 | 97 | 98 | 99 | 100 | |
|---|---|---|---|---|---|---|---|---|---|---|---|
| 1 | 91 | 92 | 93 | 94 | 95 | 96 | 97 | 98 | 99 | 100 | 1 |
| 2 | 182 | 184 | 186 | 188 | 190 | 192 | 194 | 196 | 198 | 200 | 2 |
| 3 | 273 | 276 | 279 | 282 | 285 | 288 | 291 | 294 | 297 | 300 | 3 |
| 4 | 364 | 368 | 372 | 376 | 380 | 384 | 388 | 392 | 396 | 400 | 4 |
| 5 | 455 | 460 | 465 | 470 | 475 | 480 | 485 | 490 | 495 | 500 | 5 |
| 6 | 546 | 552 | 558 | 564 | 570 | 576 | 582 | 588 | 594 | 600 | 6 |
| 7 | 637 | 644 | 651 | 658 | 665 | 672 | 679 | 686 | 693 | 700 | 7 |
| 8 | 728 | 736 | 744 | 752 | 760 | 768 | 776 | 784 | 792 | 800 | 8 |
| 9 | 819 | 828 | 837 | 846 | 855 | 864 | 873 | 882 | 891 | 900 | 9 |
| 10 | 910 | 920 | 930 | 940 | 950 | 960 | 970 | 980 | 990 | 1000 | 10 |
| 11 | 1001 | 1012 | 1023 | 1034 | 1045 | 1056 | 1067 | 1078 | 1089 | 1100 | 11 |
| 12 | 1092 | 1104 | 1116 | 1128 | 1140 | 1152 | 1164 | 1176 | 1188 | 1200 | 12 |
| 13 | 1183 | 1196 | 1209 | 1222 | 1235 | 1248 | 1261 | 1274 | 1287 | 1300 | 13 |
| 14 | 1274 | 1288 | 1302 | 1316 | 1330 | 1344 | 1358 | 1372 | 1386 | 1400 | 14 |
| 15 | 1365 | 1380 | 1395 | 1410 | 1425 | 1440 | 1455 | 1470 | 1485 | 1500 | 15 |
| 16 | 1456 | 1472 | 1488 | 1504 | 1520 | 1536 | 1552 | 1568 | 1584 | 1600 | 16 |
| 17 | 1547 | 1564 | 1581 | 1598 | 1615 | 1632 | 1649 | 1666 | 1683 | 1700 | 17 |
| 18 | 1638 | 1656 | 1674 | 1692 | 1710 | 1728 | 1746 | 1764 | 1782 | 1800 | 18 |
| 19 | 1729 | 1748 | 1767 | 1786 | 1805 | 1824 | 1843 | 1862 | 1881 | 1900 | 19 |
| 20 | 1820 | 1840 | 1860 | 1880 | 1900 | 1920 | 1940 | 1960 | 1980 | 2000 | 20 |
| 21 | 1911 | 1932 | 1953 | 1974 | 1995 | 2016 | 2037 | 2058 | 2079 | 2100 | 21 |
| 22 | 2002 | 2024 | 2046 | 2068 | 2090 | 2112 | 2134 | 2156 | 2178 | 2200 | 22 |
| 23 | 2093 | 2116 | 2139 | 2162 | 2185 | 2208 | 2231 | 2254 | 2277 | 2300 | 23 |
| 24 | 2184 | 2208 | 2232 | 2256 | 2280 | 2304 | 2328 | 2352 | 2376 | 2400 | 24 |
| 25 | 2275 | 2300 | 2325 | 2350 | 2375 | 2400 | 2425 | 2450 | 2475 | 2500 | 25 |
| 26 | 2366 | 2392 | 2418 | 2444 | 2470 | 2496 | 2522 | 2548 | 2574 | 2600 | 26 |
| 27 | 2457 | 2484 | 2511 | 2538 | 2565 | 2592 | 2619 | 2646 | 2673 | 2700 | 27 |
| 28 | 2548 | 2576 | 2604 | 2632 | 2660 | 2688 | 2716 | 2744 | 2772 | 2800 | 28 |
| 29 | 2639 | 2668 | 2697 | 2726 | 2755 | 2784 | 2813 | 2842 | 2871 | 2900 | 29 |
| 30 | 2730 | 2760 | 2790 | 2820 | 2850 | 2880 | 2910 | 2940 | 2970 | 3000 | 30 |
| 31 | 2821 | 2852 | 2883 | 2914 | 2945 | 2976 | 3007 | 3038 | 3069 | 3100 | 31 |
| 32 | 2912 | 2944 | 2976 | 3008 | 3040 | 3072 | 3104 | 3136 | 3168 | 3200 | 32 |
| 33 | 3003 | 3036 | 3069 | 3102 | 3135 | 3168 | 3201 | 3234 | 3267 | 3300 | 33 |
| 34 | 3094 | 3128 | 3162 | 3196 | 3230 | 3264 | 3298 | 3332 | 3366 | 3400 | 34 |
| 35 | 3185 | 3220 | 3255 | 3290 | 3325 | 3360 | 3395 | 3430 | 3465 | 3500 | 35 |
| 36 | 3276 | 3312 | 3348 | 3384 | 3420 | 3456 | 3492 | 3528 | 3564 | 3600 | 36 |
| 37 | 3367 | 3404 | 3441 | 3478 | 3515 | 3552 | 3589 | 3626 | 3663 | 3700 | 37 |
| 38 | 3458 | 3496 | 3534 | 3572 | 3610 | 3648 | 3686 | 3724 | 3762 | 3800 | 38 |
| 39 | 3549 | 3588 | 3627 | 3666 | 3705 | 3744 | 3783 | 3822 | 3861 | 3900 | 39 |
| 40 | 3640 | 3680 | 3720 | 3760 | 3800 | 3840 | 3880 | 3920 | 3960 | 4000 | 40 |
| 41 | 3731 | 3772 | 3813 | 3854 | 3895 | 3936 | 3977 | 4018 | 4059 | 4100 | 41 |
| 42 | 3822 | 3864 | 3906 | 3948 | 3990 | 4032 | 4074 | 4116 | 4158 | 4200 | 42 |
| 43 | 3913 | 3956 | 3999 | 4042 | 4085 | 4128 | 4171 | 4214 | 4257 | 4300 | 43 |
| 44 | 4004 | 4048 | 4092 | 4136 | 4180 | 4224 | 4268 | 4312 | 4356 | 4400 | 44 |
| 45 | 4095 | 4140 | 4185 | 4230 | 4275 | 4320 | 4365 | 4410 | 4455 | 4500 | 45 |
| 46 | 4186 | 4232 | 4278 | 4324 | 4370 | 4416 | 4462 | 4508 | 4554 | 4600 | 46 |
| 47 | 4277 | 4324 | 4371 | 4418 | 4465 | 4512 | 4559 | 4606 | 4653 | 4700 | 47 |
| 48 | 4368 | 4416 | 4464 | 4512 | 4560 | 4608 | 4656 | 4704 | 4752 | 4800 | 48 |
| 49 | 4459 | 4508 | 4557 | 4606 | 4655 | 4704 | 4753 | 4802 | 4851 | 4900 | 49 |
| 50 | 4550 | 4600 | 4650 | 4700 | 4750 | 4800 | 4850 | 4900 | 4950 | 5000 | 50 |
| | 91 | 92 | 93 | 94 | 95 | 96 | 97 | 98 | 99 | 100 | |

| | 91 | 92 | 93 | 94 | 95 | 96 | 97 | 98 | 99 | 100 | |
|---|---|---|---|---|---|---|---|---|---|---|---|
| **51** | 4641 | 4692 | 4743 | 4794 | 4845 | 4896 | 4947 | 4998 | 5049 | 5100 | **51** |
| **52** | 4732 | 4784 | 4836 | 4888 | 4940 | 4992 | 5044 | 5096 | 5148 | 5200 | **52** |
| **53** | 4823 | 4876 | 4929 | 4982 | 5035 | 5088 | 5141 | 5194 | 5247 | 5300 | **53** |
| **54** | 4914 | 4968 | 5022 | 5076 | 5130 | 5184 | 5238 | 5292 | 5346 | 5400 | **54** |
| **55** | 5005 | 5060 | 5115 | 5170 | 5225 | 5280 | 5335 | 5390 | 5445 | 5500 | **55** |
| **56** | 5096 | 5152 | 5208 | 5264 | 5320 | 5376 | 5432 | 5488 | 5544 | 5600 | **56** |
| **57** | 5187 | 5244 | 5301 | 5358 | 5415 | 5472 | 5529 | 5586 | 5643 | 5700 | **57** |
| **58** | 5278 | 5336 | 5394 | 5452 | 5510 | 5568 | 5626 | 5684 | 5742 | 5800 | **58** |
| **59** | 5369 | 5428 | 5487 | 5546 | 5605 | 5664 | 5723 | 5782 | 5841 | 5900 | **59** |
| **60** | 5460 | 5520 | 5580 | 5640 | 5700 | 5760 | 5820 | 5880 | 5940 | 6000 | **60** |
| **61** | 5551 | 5612 | 5673 | 5734 | 5795 | 5856 | 5917 | 5978 | 6039 | 6100 | **61** |
| **62** | 5642 | 5704 | 5766 | 5828 | 5890 | 5952 | 6014 | 6076 | 6138 | 6200 | **62** |
| **63** | 5733 | 5796 | 5859 | 5922 | 5985 | 6048 | 6111 | 6174 | 6237 | 6300 | **63** |
| **64** | 5824 | 5888 | 5952 | 6016 | 6080 | 6144 | 6208 | 6272 | 6336 | 6400 | **64** |
| **65** | 5915 | 5980 | 6045 | 6110 | 6175 | 6240 | 6305 | 6370 | 6435 | 6500 | **65** |
| **66** | 6006 | 6072 | 6138 | 6204 | 6270 | 6336 | 6402 | 6468 | 6534 | 6600 | **66** |
| **67** | 6097 | 6164 | 6231 | 6298 | 6365 | 6432 | 6499 | 6566 | 6633 | 6700 | **67** |
| **68** | 6188 | 6256 | 6324 | 6392 | 6460 | 6528 | 6596 | 6664 | 6732 | 6800 | **68** |
| **69** | 6279 | 6348 | 6417 | 6486 | 6555 | 6624 | 6693 | 6762 | 6831 | 6900 | **69** |
| **70** | 6370 | 6440 | 6510 | 6580 | 6650 | 6720 | 6790 | 6860 | 6930 | 7000 | **70** |
| **71** | 6461 | 6532 | 6603 | 6674 | 6745 | 6816 | 6887 | 6958 | 7029 | 7100 | **71** |
| **72** | 6552 | 6624 | 6696 | 6768 | 6840 | 6912 | 6984 | 7056 | 7128 | 7200 | **72** |
| **73** | 6643 | 6716 | 6789 | 6862 | 6935 | 7008 | 7081 | 7154 | 7227 | 7300 | **73** |
| **74** | 6734 | 6808 | 6882 | 6956 | 7030 | 7104 | 7178 | 7252 | 7326 | 7400 | **74** |
| **75** | 6825 | 6900 | 6975 | 7050 | 7125 | 7200 | 7275 | 7350 | 7425 | 7500 | **75** |
| **76** | 6916 | 6992 | 7068 | 7144 | 7220 | 7296 | 7372 | 7448 | 7524 | 7600 | **76** |
| **77** | 7007 | 7084 | 7161 | 7238 | 7315 | 7392 | 7469 | 7546 | 7623 | 7700 | **77** |
| **78** | 7098 | 7176 | 7254 | 7332 | 7410 | 7488 | 7566 | 7644 | 7722 | 7800 | **78** |
| **79** | 7189 | 7268 | 7347 | 7426 | 7505 | 7584 | 7663 | 7742 | 7821 | 7900 | **79** |
| **80** | 7280 | 7360 | 7440 | 7520 | 7600 | 7680 | 7760 | 7840 | 7920 | 8000 | **80** |
| **81** | 7371 | 7452 | 7533 | 7614 | 7695 | 7776 | 7857 | 7938 | 8019 | 8100 | **81** |
| **82** | 7462 | 7544 | 7626 | 7708 | 7790 | 7872 | 7954 | 8036 | 8118 | 8200 | **82** |
| **83** | 7553 | 7636 | 7719 | 7802 | 7885 | 7968 | 8051 | 8134 | 8217 | 8300 | **83** |
| **84** | 7644 | 7728 | 7812 | 7896 | 7980 | 8064 | 8148 | 8232 | 8316 | 8400 | **84** |
| **85** | 7735 | 7820 | 7905 | 7990 | 8075 | 8160 | 8245 | 8330 | 8415 | 8500 | **85** |
| **86** | 7826 | 7912 | 7998 | 8084 | 8170 | 8256 | 8342 | 8428 | 8514 | 8600 | **86** |
| **87** | 7917 | 8004 | 8091 | 8178 | 8265 | 8352 | 8439 | 8526 | 8613 | 8700 | **87** |
| **88** | 8008 | 8096 | 8184 | 8272 | 8360 | 8448 | 8536 | 8624 | 8712 | 8800 | **88** |
| **89** | 8099 | 8188 | 8277 | 8366 | 8455 | 8544 | 8633 | 8722 | 8811 | 8900 | **89** |
| **90** | 8190 | 8280 | 8370 | 8460 | 8550 | 8640 | 8730 | 8820 | 8910 | 9000 | **90** |
| **91** | 8281 | 8372 | 8463 | 8554 | 8645 | 8736 | 8827 | 8918 | 9009 | 9100 | **91** |
| **92** | 8372 | 8464 | 8556 | 8648 | 8740 | 8832 | 8924 | 9016 | 9108 | 9200 | **92** |
| **93** | 8463 | 8556 | 8649 | 8742 | 8835 | 8928 | 9021 | 9114 | 9207 | 9300 | **93** |
| **94** | 8554 | 8648 | 8742 | 8836 | 8930 | 9024 | 9118 | 9212 | 9306 | 9400 | **94** |
| **95** | 8645 | 8740 | 8835 | 8930 | 9025 | 9120 | 9215 | 9310 | 9405 | 9500 | **95** |
| **96** | 8736 | 8832 | 8928 | 9024 | 9120 | 9216 | 9312 | 9408 | 9504 | 9600 | **96** |
| **97** | 8827 | 8924 | 9021 | 9118 | 9215 | 9312 | 9409 | 9506 | 9603 | 9700 | **97** |
| **98** | 8918 | 9016 | 9114 | 9212 | 9310 | 9408 | 9506 | 9604 | 9702 | 9800 | **98** |
| **99** | 9009 | 9108 | 9207 | 9306 | 9405 | 9504 | 9603 | 9702 | 9801 | 9900 | **99** |
| **100** | 9100 | 9200 | 9300 | 9400 | 9500 | 9600 | 9700 | 9800 | 9900 | 10000 | **100** |
| | 91 | 92 | 93 | 94 | 95 | 96 | 97 | 98 | 99 | 100 | |

# APPENDIX II.

## A TABLE OF THE SQUARES AND SQUARE ROOTS OF THE NUMBERS FROM 1 TO 1000.

THIS table is a modification of the first part of Barlow's Tables. The advantage of this abridged table beyond its more convenient size, is that through the omission of cubes, cube roots and reciprocals, the table allows more rapid use and causes much less strain on the eyes. The latter result is furthered by giving square roots only to the third decimal instead of to the seventh.

| Num. | Square. | Squ. Root. | Num. | Square. | Squ. Root. |
|---|---|---|---|---|---|
| 1 | 1 | 1.000 | 51 | 26 01 | 7.141 |
| 2 | 4 | 1.414 | 52 | 27 04 | 7.211 |
| 3 | 9 | 1.732 | 53 | 28 09 | 7.280 |
| 4 | 16 | 2.000 | 54 | 29 16 | 7.348 |
| 5 | 25 | 2.236 | 55 | 30 25 | 7.416 |
| 6 | 36 | 2.449 | 56 | 31 36 | 7.483 |
| 7 | 49 | 2.646 | 57 | 32 49 | 7.550 |
| 8 | 64 | 2.828 | 58 | 33 64 | 7.616 |
| 9 | 81 | 3.000 | 59 | 34 81 | 7.681 |
| 10 | 1 00 | 3.162 | 60 | 36 00 | 7.746 |
| 11 | 1 21 | 3.317 | 61 | 37 21 | 7.810 |
| 12 | 1 44 | 3.464 | 62 | 38 44 | 7.874 |
| 13 | 1 69 | 3.606 | 63 | 39 69 | 7.937 |
| 14 | 1 96 | 3.742 | 64 | 40 96 | 8.000 |
| 15 | 2 25 | 3.873 | 65 | 42 25 | 8.062 |
| 16 | 2 56 | 4.000 | 66 | 43 56 | 8.124 |
| 17 | 2 89 | 4.123 | 67 | 44 89 | 8.185 |
| 18 | 3 24 | 4.243 | 68 | 46 24 | 8.246 |
| 19 | 3 61 | 4.359 | 69 | 47 61 | 8.307 |
| 20 | 4 00 | 4.472 | 70 | 49 00 | 8.367 |
| 21 | 4 41 | 4.583 | 71 | 50 41 | 8.426 |
| 22 | 4 84 | 4.690 | 72 | 51 84 | 8.485 |
| 23 | 5 29 | 4.796 | 73 | 53 29 | 8.544 |
| 24 | 5 76 | 4.899 | 74 | 54 76 | 8.602 |
| 25 | 6 25 | 5.000 | 75 | 56 25 | 8.660 |
| 26 | 6 76 | 5.099 | 76 | 57 76 | 8.718 |
| 27 | 7 29 | 5.196 | 77 | 59 29 | 8.775 |
| 28 | 7 84 | 5.292 | 78 | 60 84 | 8.832 |
| 29 | 8 41 | 5.385 | 79 | 62 41 | 8.888 |
| 30 | 9 00 | 5.477 | 80 | 64 00 | 8.944 |
| 31 | 9 61 | 5.568 | 81 | 65 61 | 9.000 |
| 32 | 10 24 | 5.657 | 82 | 67 24 | 9.055 |
| 33 | 10 89 | 5.745 | 83 | 68 89 | 9.110 |
| 34 | 11 56 | 5.831 | 84 | 70 56 | 9.165 |
| 35 | 12 25 | 5.916 | 85 | 72 25 | 9.220 |
| 36 | 12 96 | 6.000 | 86 | 73 96 | 9.274 |
| 37 | 13 69 | 6.083 | 87 | 75 69 | 9.327 |
| 38 | 14 44 | 6.164 | 88 | 77 44 | 9.381 |
| 39 | 15 21 | 6.245 | 89 | 79 21 | 9.434 |
| 40 | 16 00 | 6.325 | 90 | 81 00 | 9.487 |
| 41 | 16 81 | 6.403 | 91 | 82 81 | 9.539 |
| 42 | 17 64 | 6.481 | 92 | 84 64 | 9.592 |
| 43 | 18 49 | 6.557 | 93 | 86 49 | 9.644 |
| 44 | 19 36 | 6.633 | 94 | 88 36 | 9.695 |
| 45 | 20 25 | 6.708 | 95 | 90 25 | 9.747 |
| 46 | 21 16 | 6.782 | 96 | 92 16 | 9.798 |
| 47 | 22 09 | 6.856 | 97 | 94 09 | 9.849 |
| 48 | 23 04 | 6.928 | 98 | 96 04 | 9.899 |
| 49 | 24 01 | 7.000 | 99 | 98 01 | 9.950 |
| 50 | 25 00 | 7.071 | 100 | 1 00 00 | 10.000 |

| Num. | Square. | Squ. Root. | Num. | Square. | Squ. Root. |
|---|---|---|---|---|---|
| 101 | 1 02 01 | 10.050 | 151 | 2 28 01 | 12.288 |
| 102 | 1 04 04 | 10.100 | 152 | 2 31 04 | 12.329 |
| 103 | 1 06 09 | 10.149 | 153 | 2 34 09 | 12.369 |
| 104 | 1 08 16 | 10.198 | 154 | 2 37 16 | 12.410 |
| 105 | 1 10 25 | 10.247 | 155 | 2 40 25 | 12.450 |
| 106 | 1 12 36 | 10.296 | 156 | 2 43 36 | 12.490 |
| 107 | 1 14 49 | 10.344 | 157 | 2 46 49 | 12.530 |
| 108 | 1 16 64 | 10.392 | 158 | 2 49 64 | 12.570 |
| 109 | 1 18 81 | 10.440 | 159 | 2 52 81 | 12.610 |
| 110 | 1 21 00 | 10.488 | 160 | 2 56 00 | 12.649 |
| 111 | 1 23 21 | 10.536 | 161 | 2 59 21 | 12.689 |
| 112 | 1 25 44 | 10.583 | 162 | 2 62 44 | 12.728 |
| 113 | 1 27 69 | 10.630 | 163 | 2 65 69 | 12.767 |
| 114 | 1 29 96 | 10.677 | 164 | 2 68 96 | 12.806 |
| 115 | 1 32 25 | 10.724 | 165 | 2 72 25 | 12.845 |
| 116 | 1 34 56 | 10.770 | 166 | 2 75 56 | 12.884 |
| 117 | 1 36 89 | 10.817 | 167 | 2 78 89 | 12.923 |
| 118 | 1 39 24 | 10.863 | 168 | 2 82 24 | 12.961 |
| 119 | 1 41 61 | 10.909 | 169 | 2 85 61 | 13.000 |
| 120 | 1 44 00 | 10.954 | 170 | 2 89 00 | 13.038 |
| 121 | 1 46 41 | 11.000 | 171 | 2 92 41 | 13.077 |
| 122 | 1 48 84 | 11.045 | 172 | 2 95 84 | 13.115 |
| 123 | 1 51 29 | 11.091 | 173 | 2 99 29 | 13.153 |
| 124 | 1 53 76 | 11.136 | 174 | 3 02 76 | 13.191 |
| 125 | 1 56 25 | 11.180 | 175 | 3 06 25 | 13.229 |
| 126 | 1 58 76 | 11.225 | 176 | 3 09 76 | 13.266 |
| 127 | 1 61 29 | 11.269 | 177 | 3 13 29 | 13.304 |
| 128 | 1 63 84 | 11.314 | 178 | 3 16 84 | 13.342 |
| 129 | 1 66 41 | 11.358 | 179 | 3 20 41 | 13.379 |
| 130 | 1 69 00 | 11.402 | 180 | 3 24 00 | 13.416 |
| 131 | 1 71 61 | 11.446 | 181 | 3 27 61 | 13.454 |
| 132 | 1 74 24 | 11.489 | 182 | 3 31 24 | 13.491 |
| 133 | 1 76 89 | 11.533 | 183 | 3 34 89 | 13.528 |
| 134 | 1 79 56 | 11.576 | 184 | 3 38 56 | 13.565 |
| 135 | 1 82 25 | 11.619 | 185 | 3 42 25 | 13.601 |
| 136 | 1 84 96 | 11.662 | 186 | 3 45 96 | 13.638 |
| 137 | 1 87 69 | 11.705 | 187 | 3 49 69 | 13.675 |
| 138 | 1 90 44 | 11.747 | 188 | 3 53 44 | 13.711 |
| 139 | 1 93 21 | 11.790 | 189 | 3 57 21 | 13.748 |
| 140 | 1 96 00 | 11.832 | 190 | 3 61 00 | 13.784 |
| 141 | 1 98 81 | 11.874 | 191 | 3 64 81 | 13.820 |
| 142 | 2 01 64 | 11.916 | 192 | 3 68 64 | 13.856 |
| 143 | 2 04 49 | 11.958 | 193 | 3 72 49 | 13.892 |
| 144 | 2 07 36 | 12.000 | 194 | 3 76 36 | 13.928 |
| 145 | 2 10 25 | 12.042 | 195 | 3 80 25 | 13.964 |
| 146 | 2 13 16 | 12.083 | 196 | 3 84 16 | 14.000 |
| 147 | 2 16 09 | 12.124 | 197 | 3 88 09 | 14.036 |
| 148 | 2 19 04 | 12.166 | 198 | 3 92 04 | 14.071 |
| 149 | 2 22 01 | 12.207 | 199 | 3 96 01 | 14.107 |
| 150 | 2 25 00 | 12.247 | 200 | 4 00 00 | 14.142 |

| Nnm. | Square. | Squ. Root. | Num. | Square. | Squ. Root. |
|---|---|---|---|---|---|
| 201 | 4 04 01 | 14.177 | 251 | 6 30 01 | 15.843 |
| 202 | 4 08 04 | 14.213 | 252 | 6 35 04 | 15.875 |
| 203 | 4 12 09 | 14.248 | 253 | 6 40 09 | 15.906 |
| 204 | 4 16 16 | 14.283 | 254 | 6 45 16 | 15.937 |
| 205 | 4 20 25 | 14.318 | 255 | 6 50 25 | 15.969 |
| 206 | 4 24 36 | 14.353 | 256 | 6 55 36 | 16.000 |
| 207 | 4 28 49 | 14.387 | 257 | 6 60 49 | 16.031 |
| 208 | 4 32 64 | 14.422 | 258 | 6 65 64 | 16.062 |
| 209 | 4 36 81 | 14.457 | 259 | 6 70 81 | 16.093 |
| 210 | 4 41 00 | 14.491 | 260 | 6 76 00 | 16.125 |
| 211 | 4 45 21 | 14.526 | 261 | 6 81 21 | 16.155 |
| 212 | 4 49 44 | 14.560 | 262 | 6 86 44 | 16.186 |
| 213 | 4 53 69 | 14.595 | 263 | 6 91 69 | 16.217 |
| 214 | 4 57 96 | 14.629 | 264 | 6 96 96 | 16.248 |
| 215 | 4 62 25 | 14.663 | 265 | 7 02 25 | 16.279 |
| 216 | 4 66 56 | 14.697 | 266 | 7 07 56 | 16.310 |
| 217 | 4 70 89 | 14.731 | 267 | 7 12 89 | 16.340 |
| 218 | 4 75 24 | 14.765 | 268 | 7 18 24 | 16.371 |
| 219 | 4 79 61 | 14.799 | 269 | 7 23 61 | 16.401 |
| 220 | 4 84 00 | 14.832 | 270 | 7 29 00 | 16.432 |
| 221 | 4 88 41 | 14.866 | 271 | 7 34 41 | 16.462 |
| 222 | 4 92 84 | 14.900 | 272 | 7 39 84 | 16.492 |
| 223 | 4 97 29 | 14.933 | 273 | 7 45 29 | 16.523 |
| 224 | 5 01 76 | 14.967 | 274 | 7 50 76 | 16.553 |
| 225 | 5 06 25 | 15.000 | 275 | 7 56 25 | 16.583 |
| 226 | 5 10 76 | 15.033 | 276 | 7 61 76 | 16.613 |
| 227 | 5 15 29 | 15.067 | 277 | 7 67 29 | 16.643 |
| 228 | 5 19 84 | 15.100 | 278 | 7 72 84 | 16.673 |
| 229 | 5 24 41 | 15.133 | 279 | 7 78 41 | 16.703 |
| 230 | 5 29 00 | 15.166 | 280 | 7 84 00 | 16.733 |
| 231 | 5 33 61 | 15.199 | 281 | 7 89 61 | 16.763 |
| 232 | 5 38 24 | 15.232 | 282 | 7 95 24 | 16.793 |
| 233 | 5 42 89 | 15.264 | 283 | 8 00 89 | 16.823 |
| 234 | 5 47 56 | 15.297 | 284 | 8 06 56 | 16.852 |
| 235 | 5 52 25 | 15.330 | 285 | 8 12 25 | 16.882 |
| 236 | 5 56 96 | 15.362 | 286 | 8 17 96 | 16.912 |
| 237 | 5 61 69 | 15.395 | 287 | 8 23 69 | 16.941 |
| 238 | 5 66 44 | 15.427 | 288 | 8 29 44 | 16.971 |
| 239 | 5 71 21 | 15.460 | 289 | 8 35 21 | 17.000 |
| 240 | 5 76 00 | 15.492 | 290 | 8 41 00 | 17.029 |
| 241 | 5 80 81 | 15.524 | 291 | 8 46 81 | 17.059 |
| 242 | 5 85 64 | 15.556 | 292 | 8 52 64 | 17.088 |
| 243 | 5 90 49 | 15.588 | 293 | 8 58 49 | 17.117 |
| 244 | 5 95 36 | 15.620 | 294 | 8 64 36 | 17.146 |
| 245 | 6 00 25 | 15.652 | 295 | 8 70 25 | 17.176 |
| 246 | 6 05 16 | 15.684 | 296 | 8 76 16 | 17.205 |
| 247 | 6 10 09 | 15.716 | 297 | 8 82 09 | 17.234 |
| 248 | 6 15 04 | 15.748 | 298 | 8 88 04 | 17.263 |
| 249 | 6 20 01 | 15.780 | 299 | 8 94 01 | 17.292 |
| 250 | 6 25 00 | 15.811 | 300 | 9 00 00 | 17.321 |

| Num. | Square. | Squ. Root. |
|---|---|---|
| 301 | 9 06 01 | 17.349 |
| 302 | 9 12 04 | 17.378 |
| 303 | 9 18 09 | 17.407 |
| 304 | 9 24 16 | 17.436 |
| 305 | 9 30 25 | 17.464 |
| 306 | 9 36 36 | 17.493 |
| 307 | 9 42 49 | 17.521 |
| 308 | 9 48 64 | 17.550 |
| 309 | 9 54 81 | 17.578 |
| 310 | 9 61 00 | 17.607 |
| 311 | 9 67 21 | 17.635 |
| 312 | 9 73 44 | 17.664 |
| 313 | 9 79 69 | 17.692 |
| 314 | 9 85 96 | 17 720 |
| 315 | 9 92 25 | 17.748 |
| 316 | 9 98 56 | 17.776 |
| 317 | 10 04 89 | 17.804 |
| 318 | 10 11 24 | 17.833 |
| 319 | 10 17 61 | 17.861 |
| 320 | 10 24 00 | 17.889 |
| 321 | 10 30 41 | 17.916 |
| 322 | 10 36 84 | 17.944 |
| 323 | 10 43 29 | 17.972 |
| 324 | 10 49 76 | 18.000 |
| 325 | 10 56 25 | 18.028 |
| 326 | 10 62 76 | 18.055 |
| 327 | 10 69 29 | 18 083 |
| 328 | 10 75 84 | 18.111 |
| 329 | 10 82 41 | 18.138 |
| 330 | 10 89 00 | 18.166 |
| 331 | 10 95 61 | 18.193 |
| 332 | 11 02 24 | 18.221 |
| 333 | 11 08 89 | 18.248 |
| 334 | 11 15 56 | 18.276 |
| 335 | 11 22 25 | 18 303 |
| 336 | 11 28 96 | 18.330 |
| 337 | 11 35 69 | 18 358 |
| 338 | 11 42 44 | 18.385 |
| 339 | 11 49 21 | 18.412 |
| 340 | 11 56 00 | 18.439 |
| 341 | 11 62 81 | 18 466 |
| 342 | 11 69 64 | 18 493 |
| 343 | 11 76 49 | 18.520 |
| 344 | 11 83 36 | 18.547 |
| 345 | 11 90 25 | 18.574 |
| 346 | 11 97 16 | 18.601 |
| 347 | 12 04 09 | 18.628 |
| 348 | 12 11 04 | 18.655 |
| 349 | 12 18 01 | 18 682 |
| 350 | 12 25 00 | 18.708 |
| 351 | 12 32 01 | 18.735 |
| 352 | 12 39 04 | 18.762 |
| 353 | 12 46 09 | 18.788 |
| 354 | 12 53 16 | 18.815 |
| 355 | 12 60 25 | 18.841 |
| 356 | 12 67 36 | 18.868 |
| 357 | 12 74 49 | 18.894 |
| 358 | 12 81 64 | 18.921 |
| 359 | 12 88 81 | 18.947 |
| 360 | 12 96 00 | 18.974 |
| 361 | 13 03 21 | 19.000 |
| 362 | 13 10 44 | 19.026 |
| 363 | 13 17 69 | 19.053 |
| 364 | 13 24 96 | 19.079 |
| 365 | 13 32 25 | 19.105 |
| 366 | 13 39 56 | 19.131 |
| 367 | 13 46 89 | 19.157 |
| 368 | 13 54 24 | 19.183 |
| 369 | 13 61 61 | 19.209 |
| 370 | 13 69 00 | 19 235 |
| 371 | 13 76 41 | 19.261 |
| 372 | 13 83 84 | 19.287 |
| 373 | 13 91 29 | 19.313 |
| 374 | 13 98 76 | 19.339 |
| 375 | 14 06 25 | 19.365 |
| 376 | 14 13 76 | 19.391 |
| 377 | 14 21 29 | 19.416 |
| 378 | 14 28 84 | 19.442 |
| 379 | 14 36 41 | 19 468 |
| 380 | 14 44 00 | 19.494 |
| 381 | 14 51 61 | 19.519 |
| 382 | 14 59 24 | 19.545 |
| 383 | 14 66 89 | 19.570 |
| 384 | 14 74 56 | 19.596 |
| 385 | 14 82 25 | 19 621 |
| 386 | 14 89 96 | 19.647 |
| 387 | 14 97 69 | 19.672 |
| 388 | 15 05 44 | 19.698 |
| 389 | 15 13 21 | 19.723 |
| 390 | 15 21 00 | 19.748 |
| 391 | 15 28 81 | 19.774 |
| 392 | 15 36 64 | 19.799 |
| 393 | 15 44 49 | 19.824 |
| 394 | 15 52 36 | 19 849 |
| 395 | 15 60 25 | 19.875 |
| 396 | 15 68 16 | 19.900 |
| 397 | 15 76 09 | 19.925 |
| 398 | 15 84 04 | 19 950 |
| 399 | 15 92 01 | 19.975 |
| 400 | 16 00 00 | 20.000 |

| Num. | Square. | Squ. Root. | Num. | Square. | Squ. Root. |
|---|---|---|---|---|---|
| 401 | 16 08 01 | 20.025 | 451 | 20 34 01 | 21.237 |
| 402 | 16 16 04 | 20 050 | 452 | 20 43 04 | 21.260 |
| 403 | 16 24 09 | 20.075 | 453 | 20 52 09 | 21.284 |
| 404 | 16 32 16 | 20.100 | 454 | 20 61 16 | 21.307 |
| 405 | 16 40 25 | 20.125 | 455 | 20 70 25 | 21.331 |
| 406 | 16 48 36 | 20.149 | 456 | 20 79 36 | 21.354 |
| 407 | 16 56 49 | 20 174 | 457 | 20 88 49 | 21.378 |
| 408 | 16 64 64 | 20.199 | 458 | 20 97 64 | 21 401 |
| 409 | 16 72 81 | 20.224 | 459 | 21 06 81 | 21.424 |
| 410 | 16 81 00 | 20.248 | 460 | 21 16 00 | 21.448 |
| 411 | 16 89 21 | 20 273 | 461 | 21 25 21 | 21.471 |
| 412 | 16 97 44 | 20.298 | 462 | 21 34 44 | 21.494 |
| 413 | 17 05 69 | 20.322 | 463 | 21 43 69 | 21.517 |
| 414 | 17 13 96 | 20.347 | 464 | 21 52 96 | 21.541 |
| 415 | 17 22 25 | 20.372 | 465 | 21 62 25 | 21.564 |
| 416 | 17 30 56 | 20.396 | 466 | 21 71 56 | 21.587 |
| 417 | 17 38 89 | 20.421 | 467 | 21 80 89 | 21.610 |
| 418 | 17 47 24 | 20 445 | 468 | 21 90 24 | 21.633 |
| 419 | 17 55 61 | 20.469 | 469 | 21 99 61 | 21 656 |
| 420 | 17 64 00 | 20.494 | 470 | 22 09 00 | 21.679 |
| 421 | 17 72 41 | 20.518 | 471 | 22 18 41 | 21.703 |
| 422 | 17 80 84 | 20.543 | 472 | 22 27 84 | 21.726 |
| 423 | 17 89 29 | 20 567 | 473 | 22 37 29 | 21.749 |
| 424 | 17 97 76 | 20.591 | 474 | 22 46 76 | 21.772 |
| 425 | 18 06 25 | 20.616 | 475 | 22 56 25 | 21.794 |
| 426 | 18 14 76 | 20.640 | 476 | 22 65 76 | 21.817 |
| 427 | 18 23 29 | 20.664 | 477 | 22 75 29 | 21.840 |
| 428 | 18 31 84 | 20 688 | 478 | 22 84 84 | 21.863 |
| 429 | 18 40 41 | 20 712 | 479 | 22 94 41 | 21.886 |
| 430 | 18 49 00 | 20.736 | 480 | 23 04 00 | 21.909 |
| 431 | 18 57 61 | 20.761 | 481 | 23 13 61 | 21.932 |
| 432 | 18 66 24 | 20.785 | 482 | 23 23 24 | 21.954 |
| 433 | 18 74 89 | 20.809 | 483 | 23 32 89 | 21.977 |
| 434 | 18 83 56 | 20.833 | 484 | 23 42 56 | 22.000 |
| 435 | 18 92 25 | 20.857 | 485 | 23 52 25 | 22.023 |
| 436 | 19 00 96 | 20 881 | 486 | 23 61 96 | 22 045 |
| 437 | 19 09 69 | 20.905 | 487 | 23 71 69 | 22.068 |
| 438 | 19 18 44 | 20 928 | 488 | 23 81 44 | 22.091 |
| 439 | 19 27 21 | 20.952 | 489 | 23 91 21 | 22.113 |
| 440 | 19 36 00 | 20.976 | 490 | 24 01 00 | 22.136 |
| 441 | 19 44 81 | 21.000 | 491 | 24 10 81 | 22.159 |
| 442 | 19 53 64 | 21.024 | 492 | 24 20 64 | 22.181 |
| 443 | 19 62 49 | 21.048 | 493 | 24 30 49 | 22.204 |
| 444 | 19 71 36 | 21.071 | 494 | 24 40 36 | 22.226 |
| 445 | 19 80 25 | 21.095 | 495 | 24 50 25 | 22.249 |
| 446 | 19 89 16 | 21 119 | 496 | 24 60 16 | 22.271 |
| 447 | 19 98 09 | 21.142 | 497 | 24 70 09 | 22.293 |
| 448 | 20 07 04 | 21.166 | 498 | 24 80 04 | 22.316 |
| 449 | 20 16 01 | 21.190 | 499 | 24 90 01 | 22 338 |
| 450 | 20 25 00 | 21 213 | 500 | 25 00 00 | 22 361 |

| Num. | Square. | Squ. Root. | Num. | Square. | Squ. Root. |
|---|---|---|---|---|---|
| 501 | 25 10 01 | 22.383 | 551 | 30 36 01 | 23.473 |
| 502 | 25 20 04 | 22.405 | 552 | 30 47 04 | 23.495 |
| 503 | 25 30 09 | 22.428 | 553 | 30 58 09 | 23.516 |
| 504 | 25 40 16 | 22.450 | 554 | 30 69 16 | 23.537 |
| 505 | 25 50 25 | 22.472 | 555 | 30 80 25 | 23.558 |
| 506 | 25 60 36 | 22.494 | 556 | 30 91 36 | 23.580 |
| 507 | 25 70 49 | 22.517 | 557 | 31 02 49 | 23.601 |
| 508 | 25 80 64 | 22.539 | 558 | 31 13 64 | 23.622 |
| 509 | 25 90 81 | 22.561 | 559 | 31 24 81 | 23.643 |
| 510 | 26 01 00 | 22.583 | 560 | 31 36 00 | 23.664 |
| 511 | 26 11 21 | 22.605 | 561 | 31 47 21 | 23.685 |
| 512 | 26 21 44 | 22.627 | 562 | 31 58 44 | 23.707 |
| 513 | 26 31 69 | 22.650 | 563 | 31 69 69 | 23.728 |
| 514 | 26 41 96 | 22.672 | 564 | 31 80 96 | 23.749 |
| 515 | 26 52 25 | 22.694 | 565 | 31 92 25 | 23.770 |
| 516 | 26 62 56 | 22.716 | 566 | 32 03 56 | 23.791 |
| 517 | 26 72 89 | 22.738 | 567 | 32 14 89 | 23.812 |
| 518 | 26 83 24 | 22.760 | 568 | 32 26 24 | 23.833 |
| 519 | 26 93 61 | 22.782 | 569 | 32 37 61 | 23.854 |
| 520 | 27 04 00 | 22.804 | 570 | 32 49 00 | 23.875 |
| 521 | 27 14 41 | 22.825 | 571 | 32 60 41 | 23 896 |
| 522 | 27 24 84 | 22.847 | 572 | 32 71 84 | 23.917 |
| 523 | 27 35 29 | 22.869 | 573 | 32 83 29 | 23.937 |
| 524 | 27 45 76 | 22.891 | 574 | 32 94 76 | 23.958 |
| 525 | 27 56 25 | 22.913 | 575 | 33 06 25 | 23.979 |
| 526 | 27 66 76 | 22.935 | 576 | 33 17 76 | 24.000 |
| 527 | 27 77 29 | 22.956 | 577 | 33 29 29 | 24.021 |
| 528 | 27 87 84 | 22.978 | 578 | 33 40 84 | 24.042 |
| 529 | 27 98 41 | 23.000 | 579 | 33 52 41 | 24.062 |
| 530 | 28 09 00 | 23.022 | 580 | 33 64 00 | 24.083 |
| 531 | 28 19 61 | 23.043 | 581 | 33 75 61 | 24.104 |
| 532 | 28 30 24 | 23.065 | 582 | 33 87 24 | 24.125 |
| 533 | 28 40 89 | 23.087 | 583 | 33 98 89 | 24.145 |
| 534 | 28 51 56 | 23.108 | 584 | 34 10 56 | 24.166 |
| 535 | 28 62 25 | 23.130 | 585 | 34 22 25 | 24.187 |
| 536 | 28 72 96 | 23.152 | 586 | 34 33 96 | 24.207 |
| 537 | 28 83 69 | 23.173 | 587 | 34 45 69 | 24.228 |
| 538 | 28 94 44 | 23.195 | 528 | 34 57 44 | 24.249 |
| 539 | 29 05 21 | 23.216 | 589 | 34 69 21 | 24.269 |
| 540 | 29 16 00 | 23.238 | 590 | 34 81 00 | 24.290 |
| 541 | 29 26 81 | 23.259 | 591 | 34 92 81 | 24.310 |
| 542 | 29 37 64 | 23.281 | 592 | 35 04 64 | 24.331 |
| 543 | 29 48 49 | 23.302 | 593 | 35 16 49 | 24.352 |
| 544 | 29 59 36 | 23.324 | 594 | 35 28 36 | 24.372 |
| 545 | 29 70 25 | 23.345 | 595 | 35 40 25 | 24.393 |
| 546 | 29 81 16 | 23.367 | 596 | 35 52 16 | 24.413 |
| 547 | 29 92 09 | 23.388 | 597 | 35 64 09 | 24.434 |
| 548 | 30 03 04 | 23.409 | 598 | 35 76 04 | 24.454 |
| 549 | 30 14 01 | 23.431 | 599 | 35 88 01 | 24.474 |
| 550 | 30 25 00 | 23.452 | 600 | 36 00 00 | 24.495 |

| Num. | Square. | Squ. Root. | Num. | Square. | Squ. Root. |
|---|---|---|---|---|---|
| 601 | 36 12 01 | 24.515 | 651 | 42 38 01 | 25.515 |
| 602 | 36 24 04 | 24.536 | 652 | 42 51 04 | 25.534 |
| 603 | 36 36 09 | 24.556 | 653 | 42 64 09 | 25.554 |
| 604 | 36 48 16 | 24.576 | 654 | 42 77 16 | 25.573 |
| 605 | 36 60 25 | 24.597 | 655 | 42 90 25 | 25.593 |
| 606 | 36 72 36 | 24 617 | 656 | 43 03 36 | 25.612 |
| 607 | 36 84 49 | 24.637 | 657 | 43 16 49 | 25.632 |
| 608 | 36 96 64 | 24.658 | 658 | 43 29 64 | 25.652 |
| 609 | 37 08 81 | 24.678 | 659 | 43 42 81 | 25.671 |
| 610 | 37 21 00 | 24.698 | 660 | 43 56 00 | 25.690 |
| 611 | 37 33 21 | 24.718 | 661 | 43 69 21 | 25.710 |
| 612 | 37 45 44 | 24.739 | 662 | 43 82 44 | 25.729 |
| 613 | 37 57 69 | 24.759 | 663 | 43 95 69 | 25.749 |
| 614 | 37 69 96 | 24.779 | 664 | 44 08 96 | 25.768 |
| 615 | 37 82 25 | 24.799 | 665 | 44 22 25 | 25.788 |
| 616 | 37 94 56 | 24.819 | 666 | 44 35 56 | 25.807 |
| 617 | 38 06 89 | 24 839 | 667 | 44 48 89 | 25.826 |
| 618 | 38 19 24 | 24.860 | 668 | 44 62 24 | 25.846 |
| 619 | 38 31 61 | 24.880 | 669 | 44 75 61 | 25.865 |
| 620 | 38 44 00 | 24.900 | 670 | 44 89 00 | 25.884 |
| 621 | 38 56 41 | 24.920 | 671 | 45 02 41 | 25.904 |
| 622 | 38 68 84 | 24.940 | 672 | 45 15 84 | 25.923 |
| 623 | 38 81 29 | 24 960 | 673 | 45 29 29 | 25.942 |
| 624 | 38 93 76 | 24.980 | 674 | 45 42 76 | 25.962 |
| 625 | 39 06 25 | 25.000 | 675 | 45 56 25 | 25.981 |
| 626 | 39 18 76 | 25.020 | 676 | 45 69 76 | 26.000 |
| 627 | 39 31 29 | 25.040 | 677 | 45 83 29 | 26.019 |
| 628 | 39 43 84 | 25.060 | 678 | 45 96 84 | 26.038 |
| 629 | 39 56 41 | 25.080 | 679 | 46 10 41 | 26.058 |
| 630 | 39 69 00 | 25.100 | 680 | 46 24 00 | 26.077 |
| 631 | 39 81 61 | 25.120 | 681 | 46 37 61 | 26.096 |
| 632 | 39 94 24 | 25.140 | 682 | 46 51 24 | 26.115 |
| 633 | 40 06 89 | 25 159 | 683 | 46 64 89 | 26.134 |
| 634 | 40 19 56 | 35.179 | 684 | 46 78 56 | 26.153 |
| 635 | 40 32 25 | 25 199 | 685 | 46 92 25 | 26.173 |
| 636 | 40 44 96 | 25.219 | 686 | 47 05 96 | 26.192 |
| 637 | 40 57 69 | 25.239 | 687 | 47 19 69 | 26.211 |
| 638 | 40 70 44 | 25.259 | 688 | 47 33 44 | 26 230 |
| 639 | 40 83 21 | 25.278 | 689 | 47 47 21 | 26.249 |
| 640 | 40 96 00 | 25.298 | 690 | 47 61 00 | 26.268 |
| 641 | 41 08 81 | 25.318 | 691 | 47 74 81 | 26.287 |
| 642 | 41 21 64 | 25.338 | 692 | 47 88 64 | 26.306 |
| 643 | 41 34 49 | 25.357 | 693 | 48 02 49 | 26.325 |
| 644 | 41 47 36 | 25 3 7 | 694 | 48 16 36 | 26.344 |
| 645 | 41 60 25 | 25.3 7 | 695 | 48 30 25 | 26.363 |
| 646 | 41 73 16 | 25.417 | 696 | 48 44 16 | 26.382 |
| 647 | 41 86 09 | 25.436 | 697 | 48 58 09 | 26.401 |
| 648 | 41 99 04 | 25.456 | 698 | 48 72 04 | 26.420 |
| 649 | 42 12 01 | 25.475 | 699 | 48 86 01 | 26.439 |
| 650 | 42 25 00 | 25.495 | 700 | 49 00 00 | 26.458 |

| Num. | Square. | Squ. Root. | Num. | Square. | Squ. Root. |
|---|---|---|---|---|---|
| 701 | 49 14 01 | 26.476 | 751 | 56 40 01 | 27 404 |
| 702 | 49 28 04 | 26.495 | 752 | 56 55 04 | 27.423 |
| 703 | 49 42 09 | 26.514 | 753 | 56 70 09 | 27.441 |
| 704 | 49 56 16 | 26.533 | 754 | 56 85 16 | 27.459 |
| 705 | 49 70 25 | 26.552 | 755 | 57 00 25 | 27.477 |
| 706 | 49 84 36 | 26.571 | 756 | 57 15 36 | 27 495 |
| 707 | 49 98 49 | 26.589 | 757 | 57 30 49 | 27.514 |
| 708 | 50 12 64 | 26.608 | 758 | 57 45 64 | 27.532 |
| 709 | 50 26 81 | 26 627 | 759 | 57 60 81 | 27.550 |
| 710 | 50 41 00 | 26.646 | 760 | 57 76 00 | 27.568 |
| 711 | 50 55 21 | 26.665 | 761 | 57 91 21 | 27.586 |
| 712 | 50 69 44 | 26.683 | 762 | 58 06 44 | 27.604 |
| 713 | 50 83 69 | 26.702 | 763 | 58 21 69 | 27.622 |
| 714 | 50 97 96 | 26 721 | 764 | 58 36 96 | 27.641 |
| 715 | 51 12 25 | 26 739 | 765 | 58 52 25 | 27.659 |
| 716 | 51 26 56 | 26 758 | 766 | 58 67 56 | 27.677 |
| 717 | 51 40 89 | 26 777 | 767 | 58 82 89 | 27.695 |
| 718 | 51 55 24 | 26.796 | 768 | 58 98 24 | 27 713 |
| 719 | 51 69 61 | 26.814 | 769 | 59 13 61 | 27.731 |
| 720 | 51 84 00 | 26.833 | 770 | 59 29 00 | 27.749 |
| 721 | 51 98 41 | 26.851 | 771 | 59 44 41 | 27.767 |
| 722 | 52 12 84 | 26.870 | 772 | 59 59 84 | 27.785 |
| 723 | 52 27 29 | 26.889 | 773 | 59 75 29 | 27 803 |
| 724 | 52 41 76 | 26.907 | 774 | 59 90 76 | 27.821 |
| 725 | 52 56 25 | 26.926 | 775 | 60 06 25 | 27.839 |
| 726 | 52 70 76 | 26 944 | 776 | 60 21 76 | 27.857 |
| 727 | 52 85 29 | 26.963 | 777 | 60 37 29 | 27.875 |
| 728 | 52 99 84 | 26 981 | 778 | 60 52 84 | 27.893 |
| 729 | 53 14 41 | 27 000 | 779 | 60 68 41 | 27 911 |
| 730 | 53 29 00 | 27.019 | 780 | 60 84 00 | 27.928 |
| 731 | 53 43 61 | 27.037 | 781 | 60 99 61 | 27.946 |
| 732 | 53 58 24 | 27.055 | 782 | 61 15 24 | 27.964 |
| 733 | 53 72 89 | 27.074 | 783 | 61 30 89 | 27.982 |
| 734 | 53 87 56 | 27.092 | 784 | 61 46 56 | 28.000 |
| 735 | 54 02 25 | 27.111 | 785 | 61 62 25 | 28.018 |
| 736 | 54 16 96 | 27.129 | 786 | 61 77 96 | 28.036 |
| 737 | 54 31 69 | 27 148 | 787 | 61 93 69 | 28.054 |
| 738 | 54 46 44 | 27.166 | 788 | 62 09 44 | 28.071 |
| 739 | 54 61 21 | 27.185 | 789 | 62 25 21 | 28.089 |
| 740 | 54 76 00 | 27.203 | 790 | 62 41 00 | 28.107 |
| 741 | 54 90 81 | 27.221 | 791 | 62 56 81 | 28 125 |
| 742 | 55 05 64 | 27.240 | 792 | 62 72 64 | 28.142 |
| 743 | 55 20 49 | 27.258 | 793 | 62 88 49 | 28.160 |
| 744 | 55 35 36 | 27 276 | 794 | 63 04 36 | 28.178 |
| 745 | 55 50 25 | 27.295 | 795 | 63 20 25 | 28.196 |
| 746 | 55 65 16 | 27.313 | 796 | 63 36 16 | 28.213 |
| 747 | 55 80 09 | 27.331 | 797 | 63 52 09 | 28.231 |
| 748 | 55 95 04 | 27.350 | 798 | 63 68 04 | 28.249 |
| 749 | 56 10 01 | 27.368 | 799 | 63 84 01 | 28.267 |
| 750 | 56 25 00 | 27.386 | 800 | 64 00 00 | 28.284 |

| Num. | Square. | Squ. Root. | Num. | Square. | Squ. Root. |
|---|---|---|---|---|---|
| 801 | 64 16 01 | 28.302 | 851 | 72 42 01 | 29 172 |
| 802 | 64 32 04 | 28.320 | 852 | 72 59 04 | 29.189 |
| 803 | 64 48 09 | 28.337 | 853 | 72 76 09 | 29.206 |
| 804 | 64 64 16 | 28.355 | 854 | 72 93 16 | 29 223 |
| 805 | 64 80 25 | 28.373 | 855 | 73 10 25 | 29.240 |
| 806 | 64 96 36 | 28.390 | 856 | 73 27 36 | 29.257 |
| 807 | 65 12 49 | 28.408 | 857 | 73 44 49 | 29.275 |
| 808 | 65 28 64 | 28 425 | 858 | 73 61 64 | 29.292 |
| 809 | 65 44 81 | 28.443 | 859 | 73 78 81 | 29.309 |
| 810 | 65 61 00 | 28 460 | 860 | 73 96 00 | 29.326 |
| 811 | 65 77 21 | 28.478 | 861 | 74 13 21 | 29.343 |
| 812 | 65 93 44 | 28.496 | 862 | 74 30 44 | 29 360 |
| 813 | 66 09 69 | 28.513 | 863 | 74 47 69 | 29 377 |
| 814 | 66 25 96 | 28.531 | 864 | 74 64 96 | 29.394 |
| 815 | 66 42 25 | 28.548 | 865 | 74 82 25 | 29.411 |
| 816 | 66 58 56 | 28 566 | 866 | 74 99 56 | 29.428 |
| 817 | 66 74 89 | 28.583 | 867 | 75 16 89 | 29.445 |
| 818 | 66 91 24 | 28.601 | 868 | 75 34 24 | 29.462 |
| 819 | 67 07 61 | 28.618 | 869 | 75 51 61 | 29.479 |
| 820 | 67 24 00 | 28.636 | 870 | 75 69 00 | 29.496 |
| 821 | 67 40 41 | 28 653 | 871 | 75 86 41 | 29.513 |
| 822 | 67 56 84 | 28.671 | 872 | 76 03 84 | 29.530 |
| 823 | 67 73 29 | 28.688 | 873 | 76 21 29 | 29.547 |
| 824 | 67 89 76 | 28.705 | 874 | 76 38 76 | 29.563 |
| 825 | 68 06 25 | 28.723 | 875 | 76 56 25 | 29.580 |
| 826 | 68 22 76 | 28.740 | 876 | 76 73 76 | 29.597 |
| 827 | 68 39 29 | 28.758 | 877 | 76 91 29 | 29.614 |
| 828 | 68 55 84 | 28.775 | 878 | 77 08 84 | 29.631 |
| 829 | 68 72 41 | 28.792 | 879 | 77 26 41 | 29 648 |
| 830 | 68 89 00 | 28.810 | 880 | 77 44 00 | 29.665 |
| 831 | 69 05 61 | 28.827 | 881 | 77 61 61 | 29.682 |
| 832 | 69 22 24 | 28.844 | 882 | 77 79 24 | 29.698 |
| 833 | 69 38 89 | 28.862 | 883 | 77 96 89 | 29.715 |
| 834 | 69 55 56 | 28.879 | 884 | 78 14 56 | 29.732 |
| 835 | 69 72 25 | 28 896 | 885 | 78 32 25 | 29.749 |
| 836 | 69 88 96 | 28.914 | 886 | 78 49 96 | 29.766 |
| 837 | 70 05 69 | 28.931 | 887 | 78 67 69 | 29.783 |
| 838 | 70 22 44 | 28 948 | 888 | 78 85 44 | 29.799 |
| 839 | 70 39 21 | 28.965 | 889 | 79 03 21 | 29.816 |
| 840 | 70 56 00 | 28.983 | 890 | 79 21 00 | 29.833 |
| 841 | 70 72 81 | 29.000 | 891 | 79 38 81 | 29 850 |
| 842 | 70 89 64 | 29.017 | 892 | 79 56 64 | 29.866 |
| 843 | 71 06 49 | 29 034 | 893 | 79 74 49 | 29 883 |
| 844 | 71 23 36 | 29 052 | 894 | 79 92 36 | 29.900 |
| 845 | 71 40 25 | 29.069 | 895 | 80 10 25 | 29.916 |
| 846 | 71 57 16 | 29.086 | 896 | 80 28 16 | 29.933 |
| 847 | 71 74 09 | 29.103 | 897 | 80 46 09 | 29.950 |
| 848 | 71 91 04 | 29.120 | 898 | 80 64 04 | 29.967 |
| 849 | 72 08 01 | 29.138 | 899 | 80 82 01 | 29 983 |
| 850 | 72 25 00 | 29.155 | 900 | 81 00 00 | 30.000 |

| Num. | Square. | Squ. Root. | Num. | Square. | Squ. Root. |
|---|---|---|---|---|---|
| 901 | 81 18 01 | 30.017 | 951 | 90 44 01 | 30.838 |
| 902 | 81 36 04 | 30.033 | 952 | 90 63 04 | 30.854 |
| 903 | 81 54 09 | 30.050 | 953 | 90 82 09 | 30.871 |
| 904 | 81 72 16 | 30.067 | 954 | 91 01 16 | 30.887 |
| 905 | 81 90 25 | 30.083 | 955 | 91 20 25 | 30.903 |
| 906 | 82 08 36 | 30.100 | 956 | 91 39 36 | 30.919 |
| 907 | 82 26 49 | 30.116 | 957 | 91 58 49 | 30.935 |
| 908 | 82 44 64 | 30.133 | 958 | 91 77 64 | 30.952 |
| 909 | 82 62 81 | 30.150 | 959 | 91 96 81 | 30.968 |
| 910 | 82 81 00 | 30.166 | 960 | 92 16 00 | 30.984 |
| 911 | 82 99 21 | 30.183 | 961 | 92 35 21 | 31.000 |
| 912 | 83 17 44 | 30.199 | 962 | 92 54 44 | 31.016 |
| 913 | 83 35 69 | 30.216 | 963 | 92 73 69 | 31.032 |
| 914 | 83 53 96 | 30.232 | 964 | 92 92 96 | 31.048 |
| 915 | 83 72 25 | 30.249 | 965 | 93 12 25 | 31.064 |
| 916 | 83 90 56 | 30.265 | 966 | 93 31 56 | 31.081 |
| 917 | 84 08 89 | 30.282 | 967 | 93 50 89 | 31.097 |
| 918 | 84 27 24 | 30.299 | 968 | 93 70 24 | 31.113 |
| 919 | 84 45 61 | 30.315 | 969 | 93 89 61 | 31.129 |
| 920 | 84 64 00 | 30.332 | 970 | 94 09 00 | 31.145 |
| 921 | 84 82 41 | 30.348 | 971 | 94 28 41 | 31.161 |
| 922 | 85 00 84 | 30.364 | 972 | 94 47 84 | 31.177 |
| 923 | 85 19 29 | 30.381 | 973 | 94 67 29 | 31.193 |
| 924 | 85 37 76 | 30.397 | 974 | 94 86 76 | 31.209 |
| 925 | 85 56 25 | 30.414 | 975 | 95 06 25 | 31.225 |
| 926 | 85 74 76 | 30.430 | 976 | 95 25 76 | 31.241 |
| 927 | 85 93 29 | 30.447 | 977 | 95 45 29 | 31.257 |
| 928 | 86 11 84 | 30.463 | 978 | 95 64 84 | 31.273 |
| 929 | 86 30 41 | 30.480 | 979 | 95 84 41 | 31.289 |
| 930 | 86 49 00 | 30.496 | 980 | 96 04 00 | 31.305 |
| 931 | 86 67 61 | 30.512 | 981 | 96 23 61 | 31.321 |
| 932 | 86 86 24 | 30.529 | 982 | 96 43 24 | 31.337 |
| 933 | 87 04 89 | 30.545 | 983 | 96 62 89 | 31.353 |
| 934 | 87 23 56 | 30.561 | 984 | 96 82 56 | 31.369 |
| 935 | 87 42 25 | 30.578 | 985 | 97 02 25 | 31.385 |
| 936 | 87 60 96 | 30.594 | 986 | 97 21 96 | 31.401 |
| 937 | 87 79 69 | 30.610 | 987 | 97 41 69 | 31.417 |
| 938 | 87 98 44 | 30.627 | 988 | 97 61 44 | 31.432 |
| 939 | 88 17 21 | 30.643 | 989 | 97 81 21 | 31.448 |
| 940 | 88 36 00 | 30.659 | 990 | 98 01 00 | 31.464 |
| 941 | 88 54 81 | 30.676 | 991 | 98 20 81 | 31.480 |
| 942 | 88 73 64 | 30.692 | 992 | 98 40 64 | 31.496 |
| 943 | 88 92 49 | 30.708 | 993 | 98 60 49 | 31.512 |
| 944 | 89 11 36 | 30.725 | 994 | 98 80 36 | 31.528 |
| 945 | 89 30 25 | 30.741 | 995 | 99 00 25 | 31.544 |
| 946 | 89 49 16 | 30.757 | 996 | 99 20 16 | 31.559 |
| 947 | 89 68 09 | 30.773 | 997 | 99 40 09 | 31.575 |
| 948 | 89 87 04 | 30.790 | 998 | 99 60 04 | 31.591 |
| 949 | 90 06 01 | 30.806 | 999 | 99 80 01 | 31.607 |
| 950 | 90 25 00 | 30.822 | 1000 | 100 00 00 | 31.623 |

# APPENDIX III.

## ANSWERS TO PROBLEMS; MISCELLANEOUS PROBLEMS.

### *Answers to Problems.*

7. $1,312, since salaries between 1,000 and 1,100 are to be reckoned as averaging 1,050, and similarly for the other groups.

16.

| | Average. | A. D. | $\sigma$. | Median. | 25 percentile. | 75 percentile. |
|---|---|---|---|---|---|---|
| I. | 15.0 | 1.0 | 1.5 | 15.1 | 14.3 | 16.0 |
| II. | 163.6 | 5.8 | 7.35 | 164.0 | 159.1 | 168.3 |
| III. | 14.1 | 3.4 | 4.3 | 13.8 | 11.2 | 16.7 |

17. Case I. Av. = 40.6. A. D., $\sigma$ and P. E. from Av. = respectively 7.5, 9.6 and 6.4. Median = 40.2. A. D., $\sigma$ and P. E. from median = respectively 7.5, 9.6 and 6.4.

Case II. Av. = 98.58. A. D., $\sigma$ and P. E. from Av. = respectively .51, .68 and .41. Median = 98.61. A. D., $\sigma$ and P. E. from median = respectively .51, .68 and .41.

18. The great frequency of measures 98.0, 99.0 and 98.6 is probably due to the tendency of the observer to record even numbers and the 'normal' temperature. The two cases reported 96.0 were very likely observed simply as between 96 and 97 and then by an error recorded as 96.0. Av. = 98.58. A. D. = .53.

19. Case I. The average is 155.6; the A. D. from it of the cases above it is 18; that of the cases below it is 15. 50 per cent. of the cases above it deviate less than 12.9 from it. 50 per cent. of the cases below it deviate less than 13.3 from it. 75 per cent. of the cases above it deviate less than 25.2 from it. 75 per cent. of the cases below it deviate less than 22.1 from it. The mode is the 140–149 group. Using 145 as an approximate modal point, the A. D. from the mode of the cases above it is 20.8; that of those below it is 11.0. 50 per cent. of the cases above it deviate less than 17.0 from it. 50 per cent. of all the cases below it deviate less than 9.9 from it. 75 per cent. of the cases above it deviate less than 29.6 from it. 75 per cent. of the cases below it deviate less than 17.1 from it.

19. Case II. The average is 5.24; the A. D. from it of the cases above it is 1.2; that of those below it is .5. 60 per cent.

of the cases above deviate less than 1.0 from the average. 53.5 per cent. of the cases below deviate less than .5 from the average. The mode is 5.000; the A. D. from it of the cases above it is 1.43; that of those below it is .51. 61 per cent. of the cases above it deviate less than 1.25. 94.5 per cent. of the cases below it deviate less than .50.

20. The mode and median and P. E.'s from them and various percentile values.

21. If the form of distribution is a rectangle,

| | |
|---|---|
| $A = +1.96$ A. D. | $D = -1.22$ A. D. |
| $B = +1.48$ A. D. | $E = -1.84$ A. D. |
| $C = +\ .16$ A. D. | $F = -1.98$ A. D. |

If the form of distribution is that of the normal probability surface,

| | |
|---|---|
| $A = +3.1$ A. D. | $D = -1.1$ A. D. |
| $B = +1.5$ A. D. | $E = -2.2$ A. D. |
| $C = +\ .1$ A. D. | $F = -3.4$ A. D. |

If $A - B = B - C$ and $B - C = C - D$, etc.,

$A = +2.8$ A. D. or $+3.2$, according to the correction made.
$B = +1.5$ A. D. " $+1.7$, " " " "
$C = +\ .2$ A. D. " $+\ .2$, " " " "
$D = -1.1$ A. D. " $-1.3$, " " " "
$E = -2.4$ A. D. " $-2.8$, " " " "
$F = -3.7$ A. D. " $-4.3$, " " " "

23. (1) + 2.2.
(2) + .08.
(3) + .9.

24. Light blue — 2.28$\sigma$.
Blue-dark blue — 1.00$\sigma$.
Gray-blue green — .08$\sigma$.
Dark gray-hazel + .47$\sigma$.
Light brown-brown + .83$\sigma$.
Dark brown + 1.34$\sigma$.
Very dark brown-black + 2.16$\sigma$.

26*a*. 70 per cent.

26*b*. 35 per cent.

29. $r = +.48$.

In the answers to problems 30–42 the unreliabilities are given in terms of the $\text{P. E.}_{\text{true measure-obtained measure}}$. These can be turned into $\sigma_{\text{t.-o.}}$ and $\text{A. D.}_{\text{t.-o.}}$ by multiplying by 1.4826 and 1.1843 respectively.

30. P. $E._{t. Av.-obt. Av.}$ = .22 ; P. $E._{t. var.-obt. var.}$ = .16.
31. " .27 ; " .19.
32. " .32 ; " .22.
33. " .47 ; " .34.
34. " .16 ; " .11.
35. P. $E._{t. diff.-obt. diff.}$ = .39.
36. " = .52.
37. " = .27.
38. " = .31.
39. " = .36.
40. P. $E._{t. r.-obt. r.}$ .051.
41. " .068.
42. " .039.
43. 68.3 per cent.
44. 13.3 " "
45. .01 " "
46. 11.7 " "
47. 7.9 " "
48. 18.3 " "
49. 4.7 " "
50. 10 and 11.68 +.
51. 10 and 8.95 +.
52. 11.83 + and 8.07 –.
53. 23.9 and 10.6.
54. 9.8 and the lower limit of the distribution which will be near O.
55. 9.7 and 22.8.
56*a*. There are 124 chances out of 10,000 for it.
56*b*. " " 227 " " " " " "
56*c*. Between $A._{obt.} - 6$ and $A._{obt.} + 6$.
57*a*. There are 227 chances out of 10,000 for it.
57*b*. " " 8,664 " " " " " "
58*a*. " " 82 " " " " " "
58*b*. " " 82 " " " " " "
58*c*. " " 6,828 " " " " " "
58*d*. The chances are 20 to 1 that $A_1 - A_2$ will exceed .21 and will not exceed 2.19.
59*a*. Between .60 and .36.

59*b*. 227 out of 10,000.
60*a*. 26 out of 10,000.
60*b*. 1,160 " " "
61*a*. 6 out of 1,000.
61*b*. 975 " " "
62. 6.73 + .
63*a*. 60 out of 1,000.
63*b*. 190 " " "
63*c*. 446 " " "
63*d*. 212 " " "
63*e*. 560 " " "
64*a*. 890 out of 1,000.
64*b*. 992 " " "
64*c*. 19.7 and 12.3.
65. As high as .40, 200 chances in 1,000.
" " " .41, 46 " " "
" " " .42, 5½ " " "
" " " .50, 0 " " "
67*a*. 327 out of 1,000.
67*b*. 673 " " "
68. 28 per cent.
69. 78 " "
70. 28 " "

## MISCELLANEOUS PROBLEMS.

71. Almost any statistical study of health or crime or educational work will furnish problems in the selection of units of measure. Amongst psychological studies, those concerned with practice or fatigue or changes due to growth will be found interesting from this point of view.

72. Let the student test himself with respect to pulse, strength, reaction-time and accuracy of discrimination 40 times each, and compute from the results his central tendency and variability in each trait. He should guard against variations due to the influence of fatigue and practice.

73. Record the amount of sleep or exercise taken daily for a month or so and present the facts in form for statistical use.

74. Calculate the median, the 25 percentile and the 75 percentile for each of the traits measured in Tables VI. to XVII.

75. What is the briefest expression of the following facts that is also reasonably adequate? Cost per pupil of general school supplies (in cents) of primary departments in Manhattan and Bronx (Report of 1901): 59, 63, 64, 66, 67, 68, 69, 70, 73, 75, 75, 75, 76, 77, 77, 80, 85, 85, 86, 87, 87, 88, 88, 89, 90, 91, 91, 92, 95, 95, 96, 96, 97, 98, 99, 100, 101, 101, 101, 101, 101, 101, 102, 102, 105, 106, 107, 109, 110, 110, 110, 113, 117, 118, 120, 122, 123, 124, 124, 127, 127, 128, 130, 130, 132, 135, 142, 172. Answer, median 98.5; Q. 1.325; distribution 1, 14, 20, 19, 12, 1, 1.

In all examples that follow calculate the reliability of every result obtained, whenever the data are at hand.

76. Calculate the central tendency and variability of the following group measure:

DEATH-RATE FROM DIARRHŒA IN THIRD QUARTER.*

| Quantity. | Frequency. | Quantity. | Frequency. | Quantity. | Frequency. |
|---|---|---|---|---|---|
| 0.0 | 1 | 5.0 | 2 | 9.5 | 0 |
| 0.5 | 1 | 5.5 | 0 | 10.0 | 1 |
| 1.0 | 1 | 6.0 | 8 | 10.5 | 1 |
| 1.5 | 1 | 6.5 | 3 | 11.0 | 0 |
| 2.0 | 4 | 7.0 | 3 | 11.5 | 0 |
| 2.5 | 3 | 7.5 | 2 | 12.0 | 0 |
| 3.0 | 12 | 8.0 | 3 | 12.5 | 0 |
| 3.5 | 10 | 8.5 | 1 | 13.0 | 0 |
| 4.0 | 7 | 9.0 | 0 | 13.5 | 1 |
| 4.5 | 5 | | | | |

77. Express graphically the following group measure and calculate its central tendency and variability:

SIZE OF SCHOOLS.†

| Quantity. Number of Children in the School. | Frequency. Number of Schools. | Quantity. Number of Children in the School. | Frequency. Number of Schools. |
|---|---|---|---|
| Less than 20 | 577 | 100–199 | 131 |
| 20–29 | 821 | 200–299 | 54 |
| 30–39 | 423 | 300–399 | 38 |
| 40–49 | 239 | 400–599 | 39 |
| 50–99 | 252 | 600 or more. | 52 |

78. Calculate the central tendencies and variabilities of the two group facts given below and compare the condition in 1850 with that in 1891.

* From G. B. Langstaff, *Studies in Statistics*, p. 299.

† From the New South Wales Register of 1901.

PAUPERISM IN ENGLAND AND WALES.*

| Quantity. Per cent. of Paupers. | Frequency. Number of Registration Districts. In 1850. | In 1891. | Quantity. Per cent. of Paupers. | Frequency. Number of Registration Districts. In 1850. | In 1891. |
|---|---|---|---|---|---|
| 0.5 | 1 | | 7.5 | 44 | 1 |
| 1.0 | 4 | 18 | 8.0 | 31 | |
| 1.5 | 2 | 48 | 8.5 | 27 | 1 |
| 2.0 | 7 | 72 | 9.0 | 34 | |
| 2.5 | 11 | 89 | 9.5 | 21 | |
| 3.0 | 21 | 100 | 10.0 | 11 | |
| 3.5 | 28 | 90 | 10.5 | 12 | |
| 4.0 | 33 | 75 | 11.0 | 11 | |
| 4.5 | 46 | 60 | 11.5 | 7 | |
| 5.0 | 55 | 40 | 12.0 | 7 | |
| 5.5 | 40 | 21 | 12.5 | 3 | |
| 6.0 | 45 | 11 | 13.0 | 1 | |
| 6.5 | 44 | 5 | 13.5 | 3 | |
| 7.0 | 35 | 1 | 14.0 | 4 | |
| | | | Total = 588 | 632 | |

79. (*a*) Present graphically the table of frequency given below. (*b*) Present also the distributions which would result if selection so worked on the group that for each removal of a step from the mode 10 per cent. of the cases were eliminated, that is if only 90 per cent. of the 19's and 21's remained, only 80 per cent. of the 18's and 22's, etc. (*c*) Present also the result if 10 per cent. of the highest group were eliminated, 20 per cent. of the next highest, 30 per cent. of the next, etc. (*d*) Present also the result if there were no elimination above the mode, but below it an elimination of 3 per cent. for the nearest group, 6 per cent. for the next, 12 for the next, 24 for the next, etc. (*e*) Let the conditions be as in *d* except that the elimination be 1 per cent., 4, 9, 16, 25, etc.

A NORMAL DISTRIBUTION.

| Quantity. | Frequency. | Quantity. | Frequency. |
|---|---|---|---|
| 11 | 0.01 | 21 | 438 |
| 12 | 0.2 | 22 | 318 |
| 13 | 2 | 23 | 186 |
| 14 | 7 | 24 | 85 |
| 15 | 29 | 25 | 29 |
| 16 | 85 | 26 | 7 |
| 17 | 186 | 27 | 2 |
| 18 | 318 | 28 | 0.2 |
| 19 | 438 | 29 | 0.01 |
| 20 | 486 | | |

* From an article by G. Udny Yule in the *Journal of the Royal Statistical Society*, Vol. 59, page 347.

80. What is the evidence from the figures themselves that the form of distribution for the rate of interest given in the figures below is due to conventional rather than natural causes?

MORTGAGES ON HOMES IN NEW JERSEY.*

| Rate of Interest. | Number of Mortgages. | Amount of Mortgages (in thousands of dollars). |
|---|---|---|
| 0 | 104 | 136 |
| 1 | 5 | 6 |
| 2 | 15 | 36 |
| 3 | 64 | 130 |
| 4 | 387 | 790 |
| 5 | 10,629 | 25,430 |
| 6 | 27,956 | 38,901† |
| 7 | 428 | 407 |
| 8 | 69 | 68 |
| 9 | 30 | 33 |
| 10 | 57 | 58 |
| 11 | 8 | 7 |
| 12 | 23 | 18 |
| 13 | 2 | 1 |
| 14 | 0 | 0 |
| 15 | 6 | 3 |
| 16 | 0 | less than $500. |
| 17 | 1 | less than $500. |
| 18 | 1 | less than $500. |
| 50 | 1 | |

81. Explain why amounts of property, incomes, holdings of land, inheritances and taxes should be distributed with a mode at or near the low end and a very pronounced positive skewness, so great that the upper extreme is often 10,000 times the amount of the mode.

82. Explain the form of distribution of Fig. 87 A.

83. The modal number of children for American women married twenty years or more was, in 1700–1750, seven. It is at present two. Suppose a student of the fertility of the American race to get from a tabulation of the figures given in genealogy books the distribution of Fig. 87 B. How would you explain his result?

84. How would you explain the form of distribution of Fig. 87 C, found for the frequency of pauperism in England and Wales?

86. What would be the form of distribution of the speed of race horses?

* Taken from a report by G. H. Holmes in the *Journal of the Royal Statistical Society*, Vol. 56, p. 475.

† All but 160 at precisely 6 per cent.

87. What would be the form of distribution of the morality of criminals?

88. What would be the form of distribution of the intelligence of day-laborers compared with that of men in general?

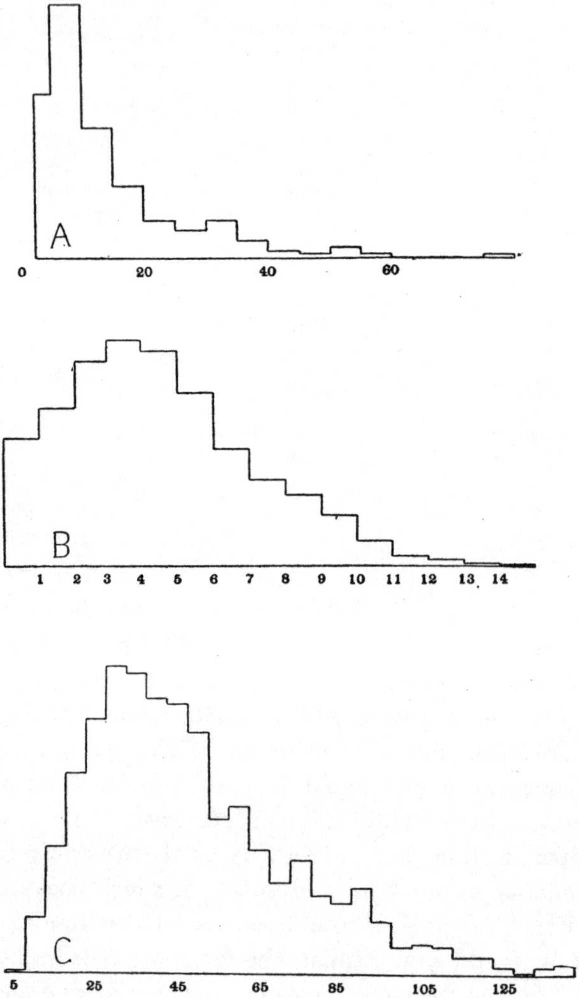


FIG. 87.

89. What would be the form of distribution of the weight of the world's war-vessels?

90. Measure the difference between boys' and girls' grammar schools with respect to the cost of supplies from the following facts:

| Cost per pupil of general supplies. In cents. | Frequency. Boys' Schools. | Frequency. Girls' Schools. | Cost per pupil of general supplies. In cents. | Frequency. Boys' Schools. | Frequency. Girls' Schools. |
|---|---|---|---|---|---|
| 100 up through 109 | | 1 | 250 | 5 | 3 |
| 110 up through 119 | | 0 | 260 | 1 | 2 |
| 120 etc. | | 2 | 270 | 2 | 0 |
| 130 | | 1 | 280 | 4 | 1 |
| 140 | 3 | 0 | 290 | 1 | 0 |
| 150 | 1 | 4 | 300 | 1 | 0 |
| 160 | 0 | 3 | 310 | 2 | 0 |
| 170 | 4 | 2 | 320 | 0 | 0 |
| 180 | 2 | 4 | 330 | 1 | 1 |
| 190 | 3 | 4 | 340 | 0 | |
| 200 | 2 | 2 | 350 | 2 | |
| 210 | 4 | 3 | | also | also |
| 220 | 2 | 2 | | 435 | 400 |
| 230 | 2 | 3 | | and | and |
| 240 | 3 | 4 | | 559 | 512 |

Answer : Gross difference, boys' schools — girls' schools = 36.6 cents if medians are compared. Only 28 per cent. of girls' schools reach the median mark for boys' schools; or 70 per cent. of boys' schools are more expensive than the median girls' school.

91. Compare the strength of pull of men with that of women, using the following facts :*

| Quantity. | Frequency in Men. | Frequency in Women. |
|---|---|---|
| 30 | 3 | 9 |
| 40 | 15 | 98 |
| 50 | 69 | 101 |
| 60 | 250 | 5 |
| 70 | 522 | 2 |
| 80 | 296 | 0 |
| 90 | 226 | 1 |
| 100 | 73 | |
| 110 | 18 | |
| 120 | 15 | |
| 130 | 2 | |
| 140 | 4 | |
| 150 | 4 | |
| Total | 1497 | 216 |

92. Compare the group of southern cities (*A*) with the group of central western cities (*B*) with respect to the regularity of attendance upon school.

* From Appendix to C. Roberts' Manual of Anthropometry.

REGULARITY OF SCHOOL ATTENDANCE.

| Percentage attendance was of enrollment. | Frequencies. *A.* | *B.* | Percentage attendance was of enrollment. | Frequencies. *A.* | *B.* |
|---|---|---|---|---|---|
| 48 | 1 | | 74 | 9 | 10 |
| 50 | | | 76 | 4 | 18 |
| 52 | 1 | | 78 | 8 | 23 |
| 54 | 1 | | 80 | 2 | 20 |
| 56 | 1 | | 82 | 1 | 12 |
| 58 | | | 84 | 1 | 6 |
| 60 | | 1 | 86 | 2 | 3 |
| 62 | 1 | 1 | 88 | | 2 |
| 64 | 5 | 1 | 90 | 1 | 1 |
| 66 | 5 | 5 | 92 | | 1 |
| 68 | 11 | 4 | 94 | | |
| 70 | 4 | 2 | 96 | | 1 |
| 72 | 5 | 3 | | 63 | 114 |

# INDEX.

Accuracy, of group measures, 42 ff.; of coefficients of correlation, 127 ff.
AIKINS, H. A., 14
Algebraic formulation of tables of frequency, 34 ff., 44
Attenuation of coefficients of correlation, 127 ff.
Averages, 32, 37 f., 41; reliability of, 139 ff.; technique of calculating, 71 ff.
Average deviation, 33, 38, 59 f.; reliability of, 142; technique of calculating, 75 f., 80

BAXTER, J. H., 46
BOWLEY, A. L., 2, 47, 70, 162

Calculation, methods of, *see* Technique
CATTELL, J. McK., 19
Causes, and measurements of relationships, 133 f.
Central tendencies, measurement of, 32, 33, 37, 38, 41; of relationships, 117, 119 f.; reliability of, 139 ff.
Changes, measurement of, 103 ff.; in a group, 106 ff.; in an individual, 104 ff.
Coefficient of correlation, 121 ff.; attenuation of, 127 ff.; reliability of, 145
COLLET, CLARA, 47
Constant errors, 157 ff.
Constants determining a table of frequency, 32, 34, 44
Correlation, 110 ff.; attenuation of, 127 ff.; Pearson coefficient of, 121 ff.; rectilinear, 119, 122
Curve of error, 36, 59, 60, 69 f.

Deviations, *see* Variability
Difference, measurements of, 97 ff., 155 f.; variability and, 98; reliability of, 142 ff.; zero points and, 98
Distribution, form of, 22 ff., 44 ff., 52, 59 f., 64, 67 f., 69 f., 147 ff.; multimodal, 31 f., 39; normal, 36, 59 f., 69 f., 147 ff.; skewed, 31 f., 38; types of, 28 ff., 44 ff.

EBBINGHAUS, H., 112, 113.
Error, curve of, 147 ff.
Errors, constant, 157 ff.; of interpretation, 159 ff.; of mean square, *see* Standard deviation; sources of, 157 ff.; variable, 157 ff.

Form of distribution, *see* Distribution
Frequency, surfaces of, 22 ff., 44; tables of, 22 ff., 44

GALE, H. G., 112
GALTON, F., 19, 96
Grades, percentile, 38, 79
Groups, changes in, 106 ff.; comparison of, 98 ff., 155 f.; measurement of, 41 ff.

HOLMES, G. H., 207
Homogeneity in groups, 53 ff.

Independence of causes and form of distribution, 64, 67 f.
Individuals, measurement of, 22 ff.; of change in, 104 ff.
Inheritance, and measurements of relationships, 133

LANGSTAFF, G. B., 205
LONDON STATISTICS, 104

Mathematics and mental measurements, 1 ff.
Measurements, by relative position, 19 ff., 39, 85 ff., 152 f.; of central tendencies, 32 ff., 139 ff.; of changes, 103 ff.; of differences 97 ff., 142 ff.; of groups, 41 ff.; of individuals, 22 ff.; of relationships, 110 ff., 145; of reliability, 136 ff.; of variability, 38, 59 f., 75 ff.
Medians, 37; calculation of, 74 f.; reliability of, 140
Mixture of species, 52 ff.
Modes, 37; calculation of, 73 f.; reliability of, 140
Multimodality, 31 f., 39, 52 f., 54 ff., 81 f.

New South Wales Register, 205
New Zealand Official Year-book, 84
Normal distribution, 36, 59, 60, 69 f., 147 ff.

Pearson coefficient of variability, 102; of correlation, 121 ff.

Pearson, K., 46, 47, 102
Percentile grades, 38, 79 ff.
Position, measurement by relative, 19 ff., 39, 85 ff., 101 f., 152 f.
Probability, theory of, 36, 59 f., 61 ff., 137 f., 143 f.
Probable error, of a distribution, 38, 59, 78 ff., 142; of a measure, 139 ff.
Problems, 21, 39 f., 82 ff., 96, 108 f., 134 f., 145 f., 151, 152, 154, 156, 204 ff.

Relationships, measurements of, 110 ff.; central tendencies of, 117, 119, 120; inheritance and, 133; reliability of, 145; variability of 110 ff.
Relative position, measurements by; *see* Position
Reliability, of measures, 136 ff., 153 f., of central tendency, 139 ff.; of correlation, 145; of difference, 142 ff.; of variability, 142
Rice, J. M., 5, 8
Roberts, C., 46, 84, 209

Scales of measurements, 15 ff.
Selection, influence of on the form of distribution, 52 f.; 57 f.
Skewness, 31 f., 38, 54 f., 57 f.; from mixture, 54 f.; from selection, 53
Spearman, C., 129
Standard deviation, of a distribution, 38, 59, 60, 76 ff., 80, 142; of a measure, 139 ff.
Subjectivity of measurements, 8 ff.
Surface of frequency, *see* Distribution

Technique of calculating measures, of change, 106 ff.; of central tendency, 71 ff.; 81 f.; of difference, 99 ff.,1 55 f.; of relationship, 110 ff.; of reliability, 139 ff., 153 f.; of transmuting measures by relative position, 85 ff.; of use of tables of frequency, 147 ff.; of variability, 75 ff.
Transmutation, of measures by position, 21, 85 ff., 152 f.

Units of measurement, 5, 7 ff., 85 ff.

Variability, 6, 22 ff., 33 f., 38, 59 f.; and the measurement of differences, 98; calculation of measures of, 75 ff.; causes of, 61 ff.; comparison of groups in, 102 f.; of a divergence as a measure of reliability, 137 ff.; of relationships, 110 ff.; reliability of measures of, 142
Variable errors, 157 ff.

Weights, 161 f.
Wilcox, W. F. 47
Wood, G. H., 24, 127.

Yule, G. U., 74, 206

Zero points, 15 f., 98, 114, 116